Dismantling the CO_2-Hoax

Preface
The aim of this text is a fundamental critique of the "theory of man-made global warming". To do this, it is not necessary to look for errors in the subtleties of complicated climate models or theories. A simple consideration of the premises of the "CO_2-theory" shows that this theory must be wrong. The most important arguments against the "CO_2-theory" are easy to understand without a background in mathematics or science. For the layman these most obvious deficiencies of the "CO_2-theory" are presented at the beginning of this text (the first 23 pages). The rest of the text is written for the amateur with some background of basic mathematics, science or engineering.
I hope that this arrangement will make this important information as accessible to the general public as possible.

May 2021, Dr. rer. nat. Markus Ott

Preface to the English Edition
Living through the madness of German climate politics, I wrote this book to offer my fellow countryman an easy to understand access to the basic science behind the man-made climate fraud. After a positive feedback from some of my peers, I decided to make the text accessible to the Anglosphere. Like the German edition this is a low budget, non-profit project, that can´t afford a professional translation or high gloss graphics. I hope my poor English and the crude presentation will not discourage the reader to engage himself in this important topic.

June 2021, Dr. rer. nat. Markus Ott

Table of contents

What or who is the driving force behind the "climate crisis" ? 2
Epistemology (Knowledge Theory): How can one know what is true and what is false? Is there an absolute truth? 4
How big is the CO_2-problem? 7
    Greenhouse gases in the Earth's atmosphere 7
Falsification of the "Theory" of Man-made Global Warming 10
    What is causing the rise in atmospheric CO_2 concentration? 11
        Reconstructing historical CO_2 data from ice cores 11
        How far back do reliable measurements of atmospheric CO_2 concentration go? 12
        Why do the high atmospheric CO_2 concentrations of the 19th century not show up in the ice core data? 14
        Reconstruction of atmospheric CO_2 concentrations via stomatal proxies 17
    Falsification of Premise 1 (man-made CO_2 increase) using official IPCC data 18
    Summary, falsification of Premise 1 (man-made CO_2 increase) 22
In advance, a few basics to refresh school knowledge 23
CO_2 Solution Equilibrium between Atmosphere and Oceans 24

Temperature dependence of the solution equilibrium 26
Cause-effect relationship between the course of temperature and atmospheric CO_2 concentration 28
Short summary of the results 31
Reconstruction of historical atmospheric temperatures via isotope ratios 32
Falsification of the greenhouse effect 34
    A few basic concepts of heat theory (thermodynamics) 34
Stefan Boltzmann Law 40
Incorrect application of the Stefan-Boltzmann law 41
Determining the greenhouse effect (according to IPCC) 42
A more realistic model for calculating the earth's average surface temperature (without greenhouse effect) .. 47
Greenhouse effect at the molecular level or "Spectroscopy of the Greenhouse effect" 53
    How to observe/measure the interaction of IR radiation with greenhouse gas molecules? 54
        What happens when CO_2 interacts with infrared light? 55
    Absorption And Emission of 15μm IR Radiation by CO2 57
    Absorption and Thermalisation of 15μm IR Radiation by CO_2 58
    Thermal Excitation and Emission of 15μm IR Radiation by CO_2 59
    "IPCC greenhouse effect" at the molecular level 60
    Saturation in the range of the 15μm radiation 62
    IPCC Trying to save the greenhouse effect 65
        Water vapour feedback 66
    Radiative transfer equation 66
    Satellite Measurements Confirm the Impermeability of the Lower Atmosphere to 15μm Radiation 69
        Stability of the excited oscillation state (01^10) of the CO_2-Molecuel 72
        Thermalisation of 15μm IR radiation in the atmosphere 72
        Note on thermally excited emission 73
        Experiments to "Proof" of the Greenhouse Effect 75
What remains of the Greenhouse effect? 75
Climate Models 76
    Climate model prediction vs. reality 76
    Fake climate data 77
Data analyses, development of models, predictions 81
Further information 87

What or who is the driving force behind the "climate crisis" ?

It is the IPCC (Intergovernmental Panel on Climate Change, https://www.ipcc.ch/). The IPCC is a political organisation of the UN, which was founded in 1988 with the aim of implementing the UN's climate policy worldwide.

The goals of this policy are derived from the UN Climate Agreement 2015 of Paris https://unfccc.int/sites/default/files/english_paris_agreement.pdf . This agreement is intended to weaken the Western industrialised nations and bring them under UN control. The industrialisation of Africa is to be prevented. No demands are made on China, the world's largest CO_2 emitter, to reduce its CO_2 emissions. On the contrary, China is being granted a generous expansion of coal-fired power production.

The IPCC is supposed to provide scientific arguments for the need of the western industrial nations to reduce their CO_2 emissions. But because there are no scientific arguments for this UN demand, thousands of "climate scientists" are bought in to give the IPCC an aura of scientific credibility. In doing so, they try to give the impression that the sponsored scientists independently arrive at results that lend weight to the IPCC's demands.

The financial support that this IPCC-compliant research receives is beyond anything imaginable. The US government's climate research budget alone from 1993 to 2014 was more than $166 billion in 2012 dollars. By comparison, the entire Apollo lunar landing programme cost about $200 billion in 2012 dollars. Under the title "Climate Dollars", the Capital Research Center has an interesting article on the financial background of climate research https://www.climatedollars.org/app/uploads/2017/05/CRC_ClimateDollars_Study_finalv3.pdf .

I know that sounds like a "conspiracy theory". But conspiracies are nothing special. They have always existed and will probably remain so.

For the IPCC staff, this situation is so self-evident that they admit in public that this is not about environmental protection (Figure 1).

Figure 1, source: https://wattsupwiththat.com/2021/04/25/wheres-the-emergency/

IPCC-sponsored climate science primarily uses climate models and manipulated climate data as instruments of deception.

With the complexity of the climate models and the absurd computer power required to run them, the following goals are being pursued.
- The aim is to create an illusion of competence among the uninformed public.
- Climate science should appear as complicated as possible, so that any attempt to understand climate events must appear hopeless to the layman.

- The enormous cost for running the climate models is intended to make climate science a monopoly of the generously sponsored IPCC scientists.

In short, the aim is to build a protective wall against any criticism of IPCC climate science.

Once you have seen through these tactics, you quickly lose respect for these specialists. Anyone who puts up such an impressive façade wants to hide something behind it or distract from something.

The IPCC wants to hide its political goals and distract us from the simple chemical and physical basics of climate phenomena. As we will see, these simple basics (mostly school/textbook knowledge) contradict the statements of the IPCC.

In the following text I will therefore not look for errors in insanely complicated climate models or discuss their manipulated data basis in detail. Although I am sure that one will find many errors and frauds there, I will only mention these things in passing and concentrate mainly on the chemical and physical basics necessary to understand the "greenhouse gas problem".

Anyone who is prepared to take a serious look at the subject is in a position to see through this fraud. The basics necessary to understand the matter are not difficult to grasp.

I have structured the text in such a way that the easiest and quickest to understand arguments are dealt with first. These easy and quick-to-understand arguments also have the charm of being the absolute "killer arguments". If you have little time and simply want to know whether climate protection measures makes any sense, you can arrive at a clear answer here with little time/reading effort (approx. 23 pages).

The UN climate policy demands dramatic interventions in our lives. We pay absurd sums for climate protection, ruin our livelihoods and are supposed to allow our freedoms to be restricted in order to save the world. In this important matter, we are advised exclusively by specialists who are directly or indirectly dependent on the IPCC.

It is high time that we do not leave this field exclusively to these specialists and activists who are leading us to ruin out of self-interest or delusion.

My aim is to make this information available to as wide an audience as possible. This PDF is therefore free of charge and may be distributed at will. I reserve all rights for any commercial use.

Before we talk about climate science, it is useful to briefly explain how science "works".

Epistemology (Knowledge Theory): How can one know what is true and what is false? Is there an absolute truth?

The clarification of these questions has been deliberately excluded from the curricula of our schools and our mainstream media do their best to blur the boundaries between true and false. A brief excursion into epistemology therefore seems useful to me.

Practising scientists often pay little attention to this aspect of their work and simply stick to the scientific methodology they were taught in their training. The scientific method is quite simple:

1. One makes an assumption.
2. One calculates the effects that this assumption has (assumption + calculated effects are then the hypothesis).
3. Tests are carried out to check whether the calculated effects can actually be observed.
4. If the experimental results do not match the calculated effects, the hypothesis is false.

5. If the experimental results confirm the hypothesis, the hypothesis is considered correct until it is refuted by an experimental result. I.e. the hypothesis must withstand constant testing by experiments.

The scientific method is therefore a method for verifying assumptions.

This methodology has been extraordinarily successful so far. Our modern world was built on this foundation. Impressed by the success of the natural sciences, perhaps also a little envious, some philosophers took a closer look at the scientific method. At the beginning of the last century they developed the philosophy of positivism from it. If you want to know more about this, you should take a look at the works of Karl Popper or A.J. Ayer (I know, this is no longer the latest thing in epistemology, but it is sufficient for our purposes). As already mentioned, you can do natural science without having read Ayer or Popper. But when it comes to judging the results of scientific work, it is very useful to have the basic concepts of this epistemology in mind. As with everything that works reliably, the basics are simple.

Statements about the world are divided into three categories:

- True statements
- False statements
- Meaningless statements

What false and meaningless statements are is quickly explained. False are statements that are logically/mathematically wrong e.g. 2 + 2 = 5 or statements that are verifiably wrong (through an experiment) e.g. "Under normal pressure water freezes at 90°C". Meaningless are statements that cannot be verified by applying logic/mathematics and defy experimental verification. A prime example of meaningless statements is the statement: "There is a teapot on the back of the moon". In principle, this statement could be tested. But because we lack the technical means to do so, we treat it as a meaningless statement.

If a statement is logically or mathematically correct, it is quite easy to recognise it as a true statement. An example of this would be 2 + 2 = 4.

It becomes more difficult when the truth of a statement has to be verified by experiments. It is in the nature of things that only a finite number of tests can be carried out to verify a statement. Even if a very large number of experiments are carried out and all of these experiments confirm the statement as true, one can never completely exclude, that one day one experiment will be carried out, whose result shows that the statement is false. If even one trial shows that the statement is false, it is to be classified as a false statement (no matter how many times it was true before).

As a result, scientific theories or hypotheses (i.e. statements that can be tested experimentally) can never be proven. They are only valid until they are disproved for the first time.

The classic example of this is the hypothesis that "all swans are white". This hypothesis held until Australia was discovered and the first black swan was seen there.

Conversely, this means that scientific theories and hypotheses must be refutable (**falsifiable**).

A good example of this is the hypothesis of man-made global warming. This hypothesis is falsifiable (refutable). To test the hypothesis, one measures the "world average temperature" for a while and then sees whether it is getting warmer or colder. The observations, temperature stays the same or temperature falls would falsify the hypothesis. This qualifies this hypothesis as a scientific hypothesis.

A few years ago, the UN changed the wording a little. Now they only speak of "man-made climate change". This means that no matter how the climate changes, the statement of this hypothesis is always true. This hypothesis cannot be falsified and is therefore not a scientific hypothesis. In such cases, one speaks of pseudoscience because one does not want to call one's colleagues frauds.

The fact that scientific theories/hypotheses are not provable has dramatic implications for the climate denier's practice. His opponents can go to any length to show that their theory/hypothesis is true. But they will never succeed in showing that their theory/hypothesis is true with absolute certainty. For the critic of the theory/hypothesis, the situation is quite different. If he finds even one error in the opposing theory/hypothesis, he has shown that this theory/hypothesis is false. His statement "the opposing theory/hypothesis is false" is then an <u>absolute truth</u>.

Against the background of this asymmetry, it becomes understandable why heretics are treated so roughly.

A single amateur can beat entire thousand-strong teams of highly paid scientists in the field of natural science.

As we will see below, the climate deniers already won the battle in the field of science decades ago. What we are currently witnessing is nothing more than an information war. Every means is being used to prevent the climate deniers from bringing the truth about the "man-made global warming" theory to the general public.

A brief explanation of the terms scientific theory vs. hypothesis: Both are statements/assumptions that can be tested by experiments. If a scientific hypothesis stands up to intensive testing and has proven itself over a longer period of time, it is elevated to the status of a theory. A theory is therefore a hypothesis on which one has been able to rely very well up to now and which one assumes will remain so. In my opinion, the best examples of theories are the theory of evolution and the laws of thermodynamics. I would be very surprised, if these theories were falsified in my lifetime. But you still can't completely rule it out.

The distinction between theory and hypothesis is often handled quite laxly. This can be seen particularly well in the term "string theory". No one has yet experimentally confirmed this nonsense and yet this construct is called a theory.

Practically every theory or hypothesis in the natural sciences is based on **premises.** A *premise* is a proposition upon which an argument is based or from which a conclusion is drawn. These premises are used in the derivation of the theory/hypothesis and are the foundation on which the theory/hypothesis is built. The theory/hypothesis can only deliver meaningful statements if these premises are fulfilled. If it turns out that even one of the premises is not fulfilled, the whole theory/hypothesis collapses.

I will illustrate this with the law of falling bodies. The formula that describes the distance travelled by objects in free fall is probably familiar (at least dimly) to everyone from school:

$s = \frac{1}{2} g t^2$ with s: distance travelled; g: Acceleration due to gravity = $9.81 m/s^2$; t: Time

In deriving this formula (or theory), the following premises were assumed:

1. The object is moving close to the earth, i.e. for all practical purposes the acceleration due to gravity acts on it at $9.81 \, m/s^2$.
2. No force other than the earth's gravity acts on the object (therefore called free fall).

A typical application of this formula is to determine the depth of a well. Everyone knows it. You throw a stone into the well and count the seconds until you hear the stone plop into the water. From the measured time, you can then calculate the well depth according to the above formula.

Premise 1 (acceleration due to gravity) is fulfilled. A stone has a fairly small surface area compared to its weight. This makes the air resistance negligible over short distances of fall. Thus, premise 2 is also fulfilled quite well and the formula provides useful results.

If you use a feather instead of a stone, this method no longer works. Apart from the fact that you don't hear it plop when the feather falls into the water, the air resistance on the feather is quite high in relation to its weight. This means that the second premise is no longer fulfilled and the application of the formula (theory) does not provide any meaningful information about the depth of the well.

Natural science is therefore not characterised by the fact that it is carried out by highly paid scientists in white lab coats, but by the fact that one works according to the scientific method.

To paraphrase Richard Feynman: "Science is the belief in the ignorance of the experts". Science is "anarchistic". Authorities count for nothing and consensus (majority opinion) does not matter. One person can show that the rest of the world is wrong.

Before we move on to the actual topic of this text, I would like to briefly explain what a scientific model is. **Scientific models** serve to illustrate scientific hypotheses or theories. When you are confronted with complicated things or processes, you often "can't see the wood for the trees". In order to bring more clarity to the situation, everything that does not seem to have any influence on the observed process is excluded from the observation. In this way, the complex reality is reduced to a minimum of objects or influences. One tries to keep the description of the process under investigation as simple as possible.

In the example mentioned above, throwing a stone into a well, the earth and the stone are described as two mass points between which only gravity acts. The entire mass of the earth is located in one of these mass points. The other mass point contains the mass of the stone. Although this model is very different from reality, it allows us to describe the free fall of a stone into a well quite well.

Scientific models can be simple mechanical arrangements, graphs, drawings, equations, computer models and much more. All scientific models have the following in common:

Model: Serves to illustrate a hypothesis/theory

    Reduces reality to simple/few influences and objects

    Good models describe the observed process "quite neatly"

    Very good models even allow predictions

In short, all models are wrong. Some models are useful.

How big is the CO_2-problem?

In addition to the epistemological basics, we should never lose sight of how big the problem is that we are discussing. Therefore, we need to get a rough idea of the scale of the "CO_2 problem".

Greenhouse gases in the Earth's atmosphere

Concentrations are often used in the context of gas mixtures. In this context, "concentration" is actually just another word for content. Concentrations can be expressed in different units of measurement. The CO_2 content of the air is about 0.04%vol and varies somewhat with the season and the location. Because 0.04%vol is an awkwardly small number, the CO_2 content is usually given in the unit ppmv. This is then 400ppmv of CO_2. ppm is an abbreviation for parts per million. The small v stands for by volume.

0.04%vol = 400ppmv

This 400ppmv of CO_2 can be visualised as follows. One cubic metre contains 1000 000ml. A cubic metre of air with 400ppmv CO_2 thus contains 400ml of CO_2-gas.

Greenhouse gases are the components of air that absorb infrared radiation (heat radiation). These are mainly water (vapour) and CO_2.

I will not go into methane here. In principle, the same applies to methane as to CO_2. It is also oxidised in the atmosphere to CO_2 and water.

The water content of the air depends strongly on the air temperature. For polar air, one calculates with approx. 0.1%vol of water vapour. Tropical air contains approx. 3%vol of water vapour. In the following I will calculate with an average value of approx. 1.3%vol water vapour (column 4, table 1).

According to the IPCC, about 3 to 4%vol of the CO_2 in the atmosphere is of human origin.

This results in the following volume fractions.

Table 1: Air composition

	Polar air Volume %	Tropical air Volume %	**Mean value Volume %**	Greenhouse gases Volume %
Nitrogen	77,303%	75,741%	77,069%	
Oxygen	20,737%	20,318%	20,674%	
Argon	0,925%	0,906%	0,922%	
Water	0,100%	3,000%	**1,300%**	96,938%
Natural CO2	0,040%	0,039%	0,039%	2,944%
"Human" CO2	0,002%	0,002%	0,002%	**0,118%**
Greenhouse gases	0,142%	3,041%	1,341%	100,000%

According to this, the "average air" contains approx. 1.3%vol of greenhouse gases. This is mainly water vapour (approx. 1.3%vol). The CO_2 share is about 0.04%vol. If we calculate with the human share of 4% of the total CO2 estimated by the IPCC, this results in a little more than 0.1%vol for the share of atmospheric greenhouse gases caused by humans.

It is not easy to represent this tiny "human" share of CO_2 in a graph. For lack of better options, I will try a "nested" pie chart (Figure 2).

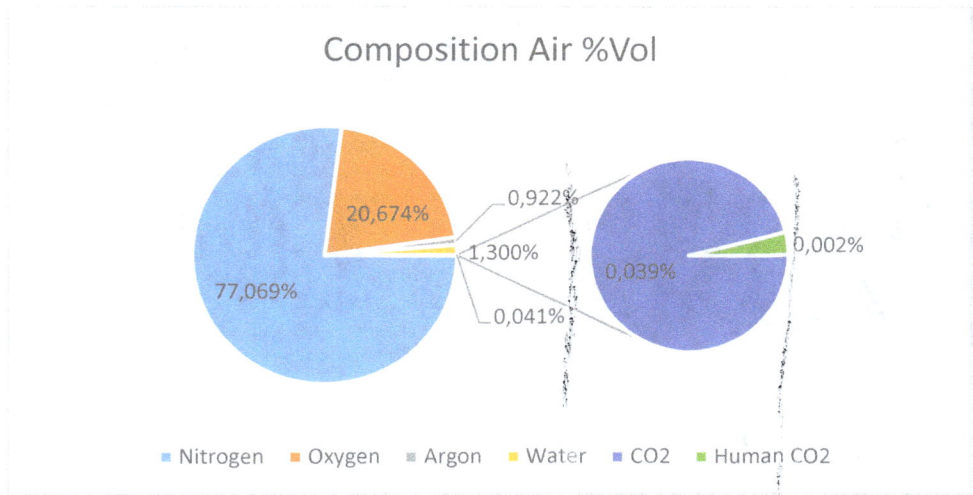

Figure 2

I.e. water vapour is the absolutely dominant greenhouse gas (95%Vol to 97%Vol of total greenhouse gases). CO_2 makes up approx. 3%vol to 4%vol of the greenhouse gases. Of this 3%vol to 4%vol, the share caused by humans is said to be approx. 4%vol, i.e. only slightly more than 0.1%vol of the total greenhouse gas quantity.

Again, **about one-thousandth (1/1000th) of greenhouse gases are thought to be of human origin, according to the IPCC (Figure 3).**

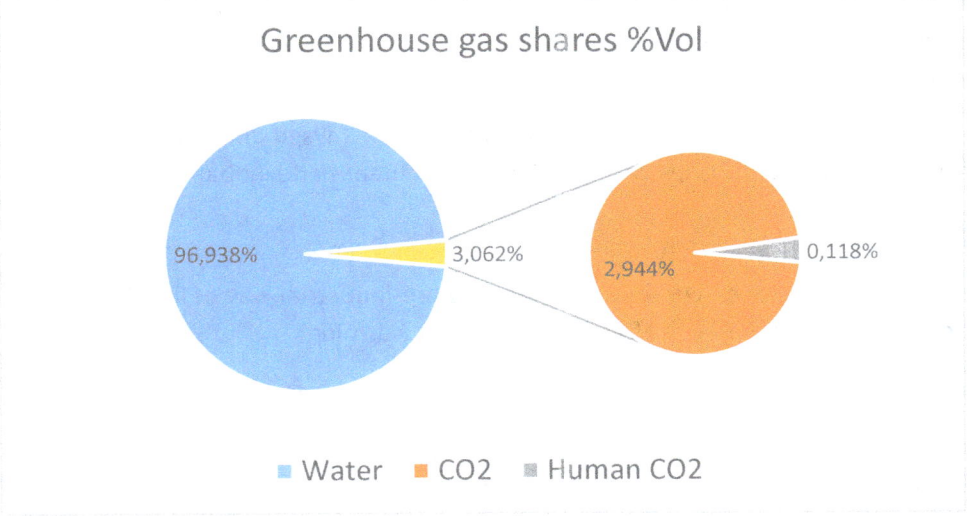

Figure 3

The fact that I am supposed to worry about this thousandth (1/1000th) while the most important greenhouse gas (water vapour) can vary by a factor of 30 depending on the weather is remarkable.

Now that we have an idea of how small the share of greenhouse gases of human origin is in the atmospheric greenhouse gases, let's take a look at the size ratio between our atmosphere and our heat source, the sun.

Figure 4: Source: https://rense.com/1.imagesH/13db967.jpg

The Sun is about 1.3 million times the size of the Earth and about 150 000 000km away from us. Our Earth is surrounded by an atmosphere that is about 0.016 Earth radii thick (Figure 4).

This thin envelope of air is said to contain approx. 0.002%vol to 0.006%vol of CO_2 of human origin. Modern, UN-sponsored climate science now wants to explain to us that this tiny proportion of a harmless gas determines the world's climate while the sun, which is about 1.3 million times the size of the earth, should have no significant influence.

Now that the reader has an idea of the scale of the problem and I made myself guilty of the worst polemics and trivialisation, we can get to grips with the "theory" of man-made global warming.

Falsification of the "Theory" of Man-made Global Warming

The UN demands that "to save the world" we should return to a pre-industrial way of life. Climate science supports this demand with the theory of man-made global warming.

The theory of man-made global warming is based on the following premises.

1. **The steady increase of CO_2 levels in the atmosphere, observed since the beginning of the Industrial Revolution, is caused by humans burning fossil fuels.**
2. **There is an atmospheric greenhouse effect.** The increase in CO_2 concentration in the atmosphere intensifies the atmospheric greenhouse effect and thus leads to a (dangerous) warming of the earth.

These premises are absolutely indispensable for the IPCC's "Man-made Global Warming" theory. **If even one of these premises can be disproved, the whole climate crisis will be exposed as a fallacy or fraud.**

Essentially, I will limit myself to showing that these two fundamental premises are not fulfilled.

What is causing the rise in atmospheric CO_2 concentration?

Reconstructing historical CO_2 data from ice cores

In order to show that the premise 1 (man causes CO_2 increase) is fulfilled, the IPCC repeatedly comes up with a reconstruction of historical atmospheric CO_2 concentrations.

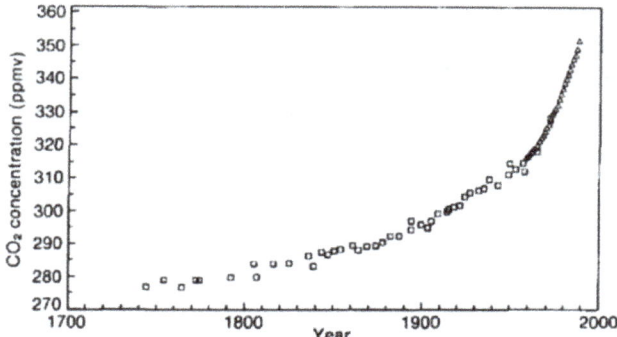

Figure 1.3: Atmospheric CO_2 increase in the past 250 years, as indicated by measurements on air trapped in ice from Siple Station, Antarctica (squares, Neftel et al., 1985a, Friedli et al, 1986) and by direct atmospheric measurements at Mauna Loa, Hawaii (triangles, Keeling et al, 1989a)

Figure 5: Source "Climate Change the IPCC Scientific Assesment" (1990)

Figure 5 is taken from the IPPC report of 1990. The vertical axis of the graph shows the CO_2 concentration in the atmosphere, measured in ppmv. The horizontal axis shows the year to which the respective measured value is assigned (time axis). CO_2 values measured in the atmosphere (measuring station on Mauna Loa, Hawaii) have been entered since 1958. The values before 1958 were obtained from the first Antarctic deep ice drilling expedition.

In this reconstruction of historical atmospheric CO_2 concentrations, IPCC science assumes that airbubbles are trapped in the ice during the formation of the glaciers. In order to access the very old ice and the air trapped in it, deep boreholes are being drilled on glaciers in Greenland and Antarctica at enormous expense. From the ice cores obtained in this way, the trapped air can be released and its CO_2 content determined. This is done under the assumption, that air is perfectly preserved in the glacier ice and that its CO_2 content does not change over the centuries. The age of ice core samples is usually determined by the stratification visible in the ice (<= change of seasons) and the depth of sampling.

Essential for the IPCC's argumentation is, that exclusively very low CO_2 concentrations for the pre-industrial period are found in the ice cores. The famous and repeatedly cited pre-industrial atmospheric CO_2 content of 280ppmv is the most important cornerstone of UN-IPCC climate science.

One needs this low pre-industrial CO_2 content to make today´s 400ppmv appear dramatic.

The air bubbles trapped in the ice cores are under very high pressure (up to approx. 350bar). Sampling (core drilling) is not a very gentle process. When the cores are taken from great depths, the pressure on the cores changes dramatically (from several hundred bar to atmospheric pressure). This creates cracks, into which the drilling water can penetrate. The assumption that the CO_2 content of the trapped air does not change under these conditions is certainly wrong. Zbigniew Jaworowski (2007) deals with this problem in detail.

Zbigniew Jaworowski: "CO2: The Greatest Scientific Scandal Of Our Time", (21st Century Science & Technology 2007), https://21sci-tech.com/Articles%202007/20_1-2_CO2_Scandal.pdf The article is well worth reading.

The use of these icecore data is particularly suspicious in view of the fact that there are other, far less costly, methods for reconstructing historical atmospheric CO_2 concentrations.

How far back do reliable measurements of atmospheric CO_2 concentration go?

The most reliable and cheapest source of historical CO_2 data are old records of CO_2 laboratory measurements.   Very reliable methods for analysing the CO_2 content of the air have existed since the beginning of the 19th century.

Ernst-Georg Beck has compiled over 90,000 historical measurements of atmospheric CO_2 and reviewed the associated documentation. The measurements cover the period from 1812 to 1961 and were taken at special measuring stations in the northern hemisphere. These measurements were carried out by reputable scientists. The measurement errors of the methods used are mostly below 3%. https://www.friendsofscience.org/assets/files/documents/CO2%20Gas%20Analysis-Ernst-Georg%20Beck.pdf

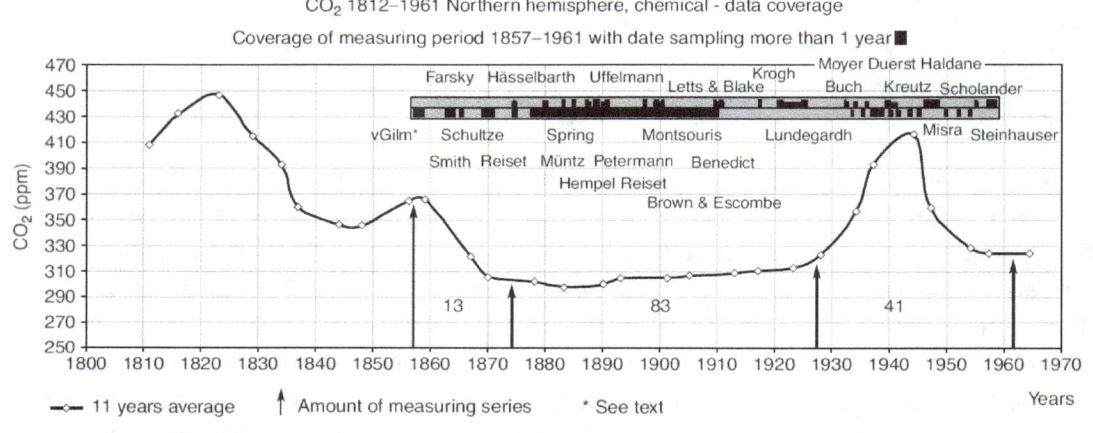

Figure 11: Local CO_2 concentration for the northern hemisphere, determined through chemical analysis between 1812 and 1861. Data plotted as an 11 year average. Data coverage and important scientists indicated in dark grey/black. The curve delineates three major maxima in CO_2 content, though the one situated around 1820 must be treated as provisional only. Data series used: time window 1857–1873: 13 yearly averages, 83 until 1927 and up to 1961 41 data records (eleven interpolated).

Figure 6: Analytical results of atmospheric CO2 compiled for the period from 1812 to 1961 (Beck 2007)

The records clearly show that in the 1820s and 1940s atmospheric CO_2 concentrations were higher than today's 400ppmv. Figure 6 shows a compilation of these data. **Premise 1 (pre-industrial 280ppmv CO_2) of the IPCC´s man-made global warming theory is thus invalid beyond doubt.** From a scientific point of view, this puts an end to the climate debate. What we are currently experiencing is not a scientific debate, but an information war only.

180 YEARS OF ATMOSPHERIC CO_2 GAS ANALYSIS BY CHEMICAL METHODS

Ernst-Georg Beck

Dipl. Biol. Ernst-Georg Beck, 31 Rue du Giessen, F-68600 Biesheim, France
E-mail: egbeck@biokurs.de; 2/2007

ABSTRACT

More than 90,000 accurate chemical analyses of CO_2 in air since 1812 are summarised. The historic chemical data reveal that changes in CO_2 track changes in temperature, and therefore climate in contrast to the simple, monotonically increasing CO_2 trend depicted in the post-1990 literature on climate-change. Since 1812, the CO_2 concentration in northern hemispheric air has fluctuated exhibiting three high level maxima around 1825, 1857 and 1942 the latter showing more than 400 ppm.

Between 1857 and 1958, the Pettenkofer process was the standard analytical method for determining atmospheric carbon dioxide levels, and usually achieved an accuracy better than 3%. These determinations were made by several scientists of Nobel Prize level distinction. Following Callendar (1938), modern climatologists have generally ignored the historic determinations of CO_2, despite the techniques being standard text book procedures in several different disciplines. Chemical methods were discredited as unreliable choosing only few which fit the assumption of a climate CO_2 connection.

Figure 7: Abstract of Beck's article, https://www.friendsofscience.org/assets/files/documents/CO2%20Gas%20Analysis-Ernst-Georg%20Beck.pdf

You can see from the historical CO₂ meters shown below (Figure 8) that these measurement methods were standardised routine methods and not unreliable, occasional laboratory experiments as the IPCC would have us believe. The methods are so accurate that weather fluctuations, changes in seasons and even changes in the phases of the moon can be detected in the regularly recorded data.

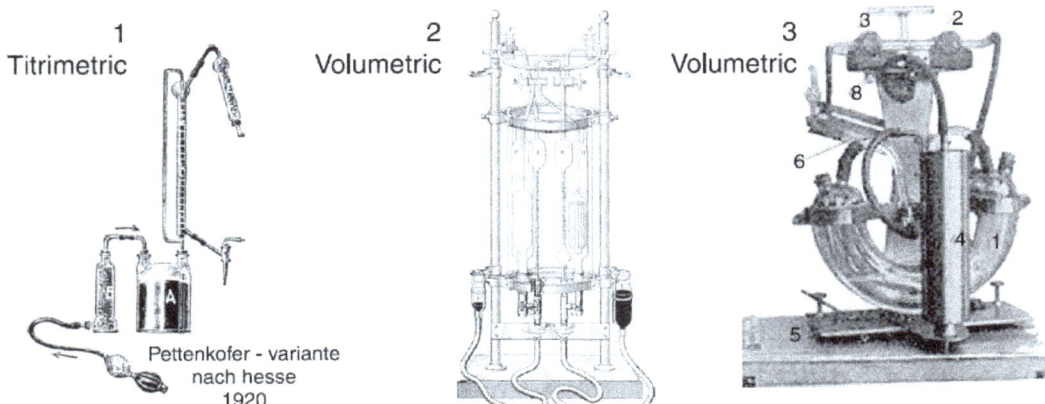

Figure 3: Important historic gas analysers used by hundreds of scientists up to 1961 [26, 42, 43, 44].

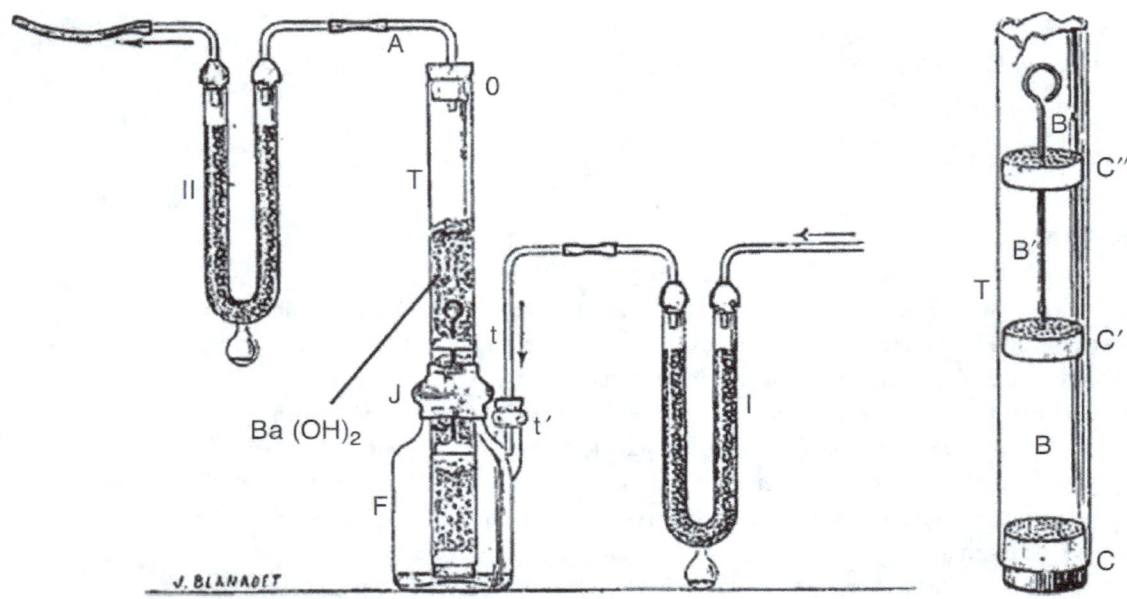

Figure 4: Part of equipment used by Reiset at Dieppe (F) 1872–80 with sulfuric acid for drying air (31). I = U-tube with sulfuric acid.

Figure 8: Historical CO_2 measuring devices, source: Beck

Why do the high atmospheric CO_2 concentrations of the 19th century not show up in the ice core data?

The answer to this question is amazingly simple. One has cheated.

Let's have a look at the original data from the first Antarctic glacier deep drilling (Siple Station west Antarctica, Antarctic summer 1983-84) (Figure 9). The data of this glacier drilling campaign are still available to everyone on the web page of the "Carbon Dioxide Information Analysis Centre CDIAC".

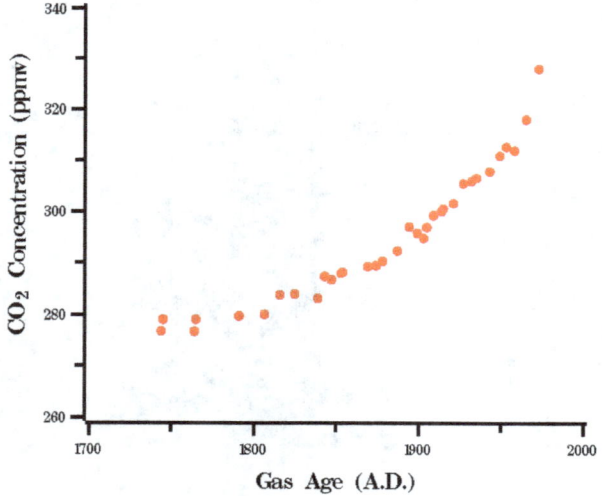

Figure 9: Source: CDIAC, Siple Station CO_2 ice core data as a function of time, https://cdiac.ess-dive.lbl.gov/trends/co2/graphics/siple-gr.gif

The graphical representation of the measurement results is unremarkable. The data seem to confirm the official story of man-made atmospheric CO_2 increase. With the beginning of the industrial revolution (approx. 1750), CO_2 concentrations, starting from pre-industrial 280ppmv, rise steadily up to the present.

Here is a copy of the original dataset published by the CDIAC:

```
************************************************************************
Historical CO2 Record from the Siple Station Ice Core ***
***                                                              ***
*** September 1994                                               ***
***                                                                    ***
*** Source: A. Neftel ***
*** H. Friedli ***
*** E. Moor ***
*** H. Lotscher ***
*** H. Oeschger ***
*** U. Siegenthaler ***
*** B. Stauffer ***
*** Physics Institute ***
*** University of Bern ***
CH-3012 Bern, Switzerland ***
************************************************************************

                                                    CO2
                    Date of  Date air  concentration in
  Depth  Samples   ice enclosed  extracted air
   (m)   measured  (yr AD)  (yr AD)   (ppmv)
187.0-187.3   10     1663    1734-1756    279
177.0-177.3   10     1683    1754-1776    279
162.0-162.3    9     1723    1794-1819    280
147.0-147.2   10     1743    1814-1836    284
128.0-129.0   47     1782    1842-1864    288
111.0-112.0   26     1812    1883-1905    297
102.0-103.0   26     1832    1903-1925    300
 92.0-93.0    25     1850    1921-1943    306
 82.0-83.0    28     1867    1938-1960    311
 76.2-76.6    11     1876    1947-1969    312
 72.4-72.7    11     1883    1954-1976    318
 68.2-68.6     8     1891    1962-1983    328
************************************************************************
Average              CO2
  depth     Gas    concentration
   (m)    (yr AD)   (ppmv)
187.70    1744     276.8
177.50    1764     276.7
168.30    1791     279.7
154.89    1816     283.8
142.75    1839     283.1
140.75    1843     287.4
138.20    1847     286.8
134.47    1854     288.2
126.80    1869     289.3
123.80    1874     289.5
121.80    1878     290.3
116.82    1887     292.3
110.20    1899     295.8
108.80    1903     294.8
107.20    1905     296.9
105.25    1909     299.2
101.80    1915     300.5
 98.80    1921     301.6
 95.17    1927     305.5
 90.77    1935     306.6
 86.80    1943     307.9
 81.22    1953     312.7

Data in the first table were published in Neftel et al. (1985); data in the second
table were published by Friedli et al. (1986).

CO2 concentrations are expressed in parts per million by volume.
```

Source: https://cdiac.ess-dive.lbl.gov/trends/co2/siple.html , https://cdiac.ess-dive.lbl.gov/ftp/trends/co2/siple2.013

Two versions of the data are given, the second version (by Friedli et al. 1986) corresponds to the graphical representation of the data shown above (Figure 9).

Much more interesting is the first version. This table, published by Neftel et al. in 1985, holds a small revelation. In the ice deposited on the glacier around the year **1891**, CO_2 concentrations of **328 ppmv** are found. This value does not at all match the pre-industrial 280ppmv called for by the IPCC. **Such a high value should actually only be reached in the 1980s.**

These high CO_2 concentrations in the upper ice layers pose a serious problem for the IPCC scientists. **If they let these values stand uncorrected, they absolutely do not fit the sacrosanct pre-industrial 280ppmv.** Therefore, to save the theory of human-caused global warming, the values must somehow be adjusted. If they adjust the values, they admit that CO_2 data from ice cores are not suitable for producing reasonably accurate reconstructions of historical atmospheric CO_2 concentrations.

To get out of the fix, the researchers fall for a simple trick. They claim that these readings can only be so high because the ice is still exchanging gases with the atmosphere in the first eight decades after it was deposited. But after that, the composition of the trapped air no longer changes for millennia. Under this pretext, they shift the time axis of their measurements until the results from the ice fit the CO_2 measurements from Mauna Loa (Hawaii) (Figure 10).

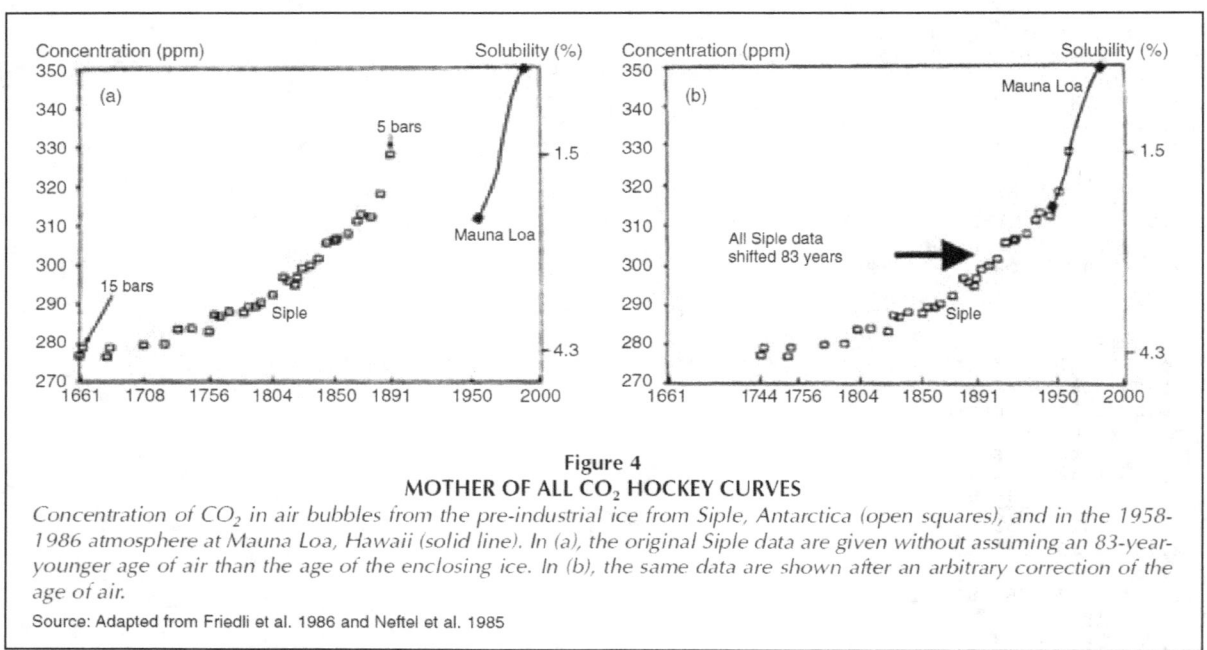

Figure 4
MOTHER OF ALL CO₂ HOCKEY CURVES

Concentration of CO₂ in air bubbles from the pre-industrial ice from Siple, Antarctica (open squares), and in the 1958-1986 atmosphere at Mauna Loa, Hawaii (solid line). In (a), the original Siple data are given without assuming an 83-year-younger age of air than the age of the enclosing ice. In (b), the same data are shown after an arbitrary correction of the age of air.
Source: Adapted from Friedli et al. 1986 and Neftel et al. 1985

Figure 10: Source: Zbigniew Jaworowski: To save the hypothesis of man-made global warming, simply shift the time axis of the measurements by 83 years.

Another way to determine CO_2 concentrations in historical atmospheres is to measure the so-called stomatal density of historical or fossil plant material.

Reconstruction of atmospheric CO_2 concentrations via stomatal proxies

Plants absorb CO_2 from the air. In photosynthesis, they build new plant material from it. They cannot absorb the CO_2 directly through their outer skin. Therefore, they have special stomata on their leaves (Figure 11) through which the gas exchange with the environment is accomplished.

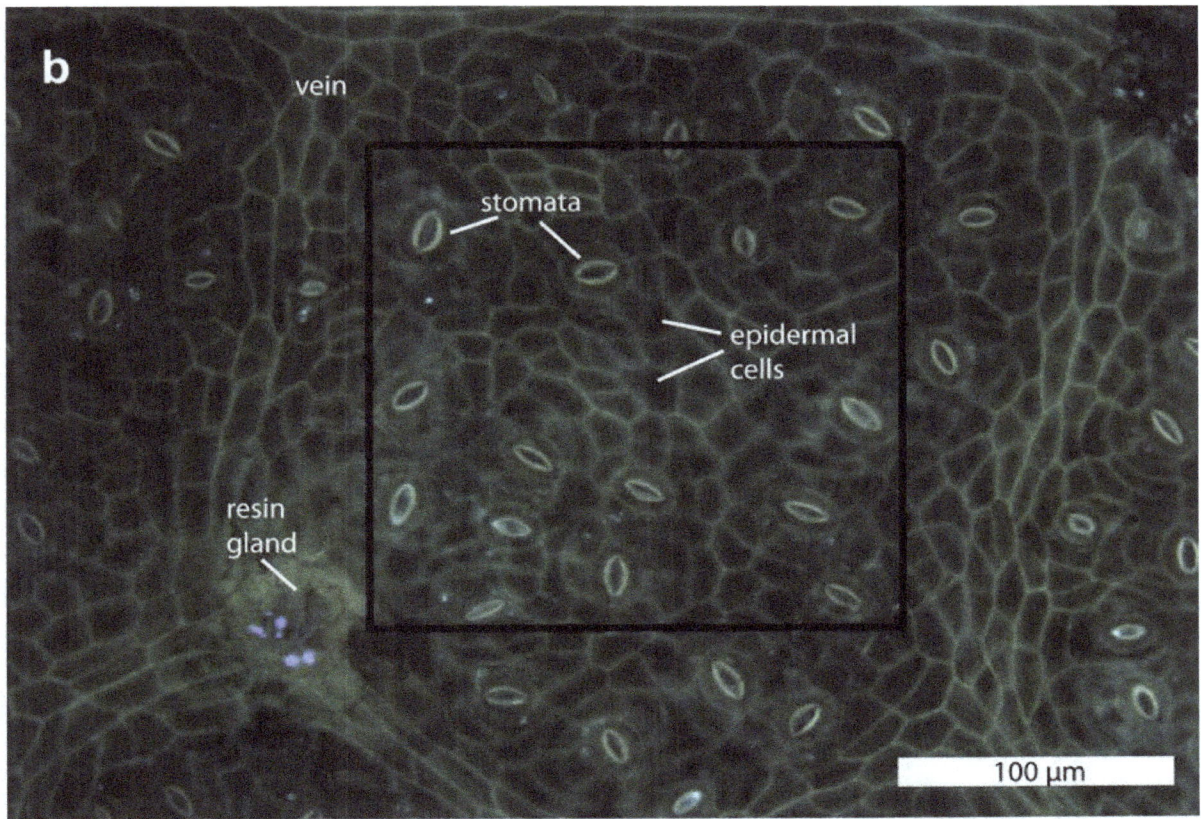

Figure 11: Fission cells (Somata) Source: Stomatal proxy record of CO2 concentrations from the last termination suggests an important role for CO2 at climate change transitions Margret Steinthorsdottir et al. Quaternary Science Reviews 68 (2013) 43-58, http://people.geo.su.se/barbara/pdf/Steinthorsdottir%20et%20al%202013%20QSR.pdf

Since plants always lose water through their stomata when they absorb CO_2, they try to keep the number of stomata as low as possible. Plants that grow in an atmosphere with a high CO_2 content therefore form fewer stomata in relation to the leaf area than plants that grow in an atmosphere with a low CO_2 content.

It turns out, that the stomatal density (number of stomatal cells per area) of leaves within a plant species, correlates well with the CO_2 concentration at which the plants grew. Thus the stomatal density of historical or fossil leaf material allows for the reconstruction of the CO_2 contents of past atmospheres.

The following figure (Figure 12) shows a comparison between a course of CO_2 concentration determined via stomatal densities and corresponding ice core data over the last 14000 years. It can be seen that the ice core data do not react to short-term changes in CO_2 concentration and are always significantly below the "stomatal data". **It is interesting that in the reconstruction via stomatal densities 400ppmv atmospheric CO_2, even in very old samples, are nothing unusual.**

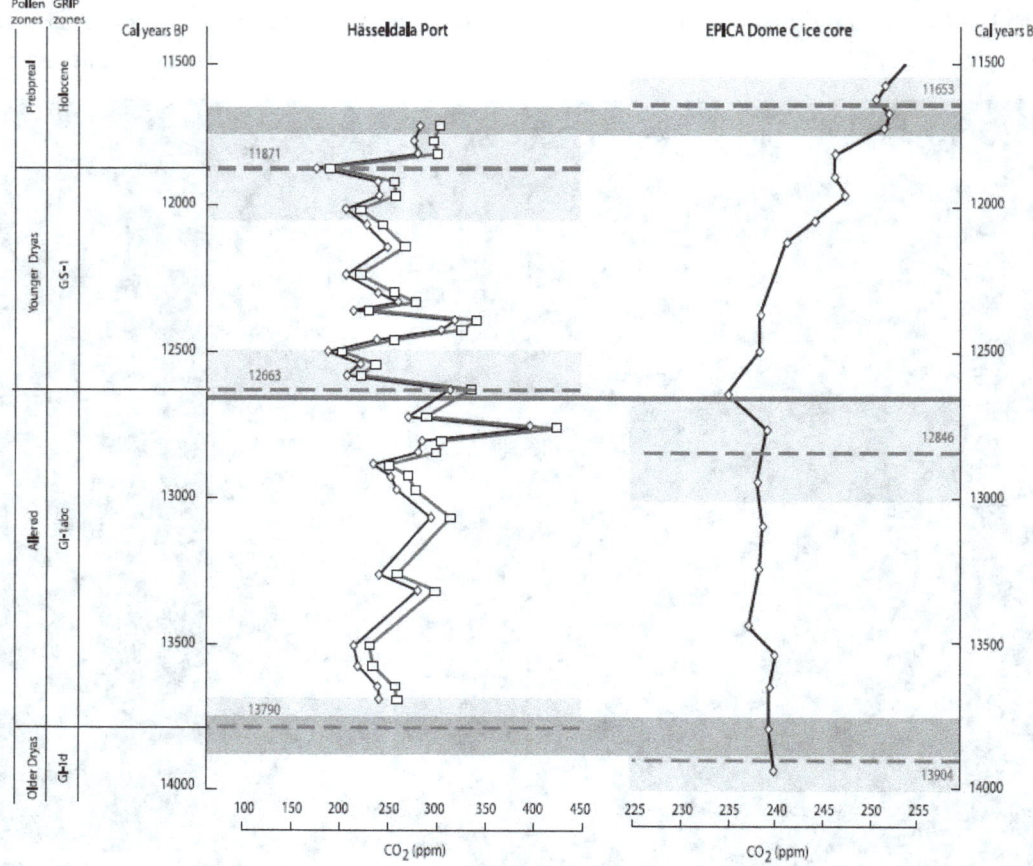

Fig. 8. Comparison with Antarctic ice core-based atmospheric CO_2 record. On the left hand side is the Hässeldala Core 5 stomatal index based CO_2 concentration record, showing approximate minimum and maximum CO_2 concentrations (see Table 3 for errors). On the right hand side is an Antarctic CO_2 record reconstructed from air bubbles in the ice cores obtained at Dome C by the EPICA project (Monnin et al., 2001), synchronized with the Greenland ice core timescale (Lemieux-Dudon et al., 2010). The ages of the main climate change boundaries for each record are illustrated with dashed lines, surrounded by their error ranges in light grey (based on Walker et al., 2008 for the Greenland ice core chronology). The darker grey bars, which overlap the error ranges shown in light grey, illustrate that the ages for each of the boundaries are comparable within their error ranges. The records, although displaying some similarities, are clearly different. Firstly, the magnitude and range of CO_2 concentrations are much larger in the Hässeldala Port record. Secondly, the stomatal-based record shows a more dynamic CO_2 development through time, in particular across the climate change boundaries, while the ice core-based record shows an almost linear, smoothed development.

Figure 12: Comparison of CO2 concentrations on the left reconstructed via stomatal density/index on the right from ice core. Source: Stomatal proxy record of CO2 concentrations from the last termination suggests an important role for CO2 at climate change transitions Margret Steinthorsdottir et al. Quaternary Science Reviews 68 (2013) 43-58, http://people.geo.su.se/barbara/pdf/Steinthorsdottir%20et%20al%202013%20QSR.pdf

What has been said here is not to say that climate data obtained from ice cores is completely useless. These data have contributed significantly to the understanding of climate events on this planet. But especially in the case of CO_2, a gas that is very soluble in water and very mobile, ice core data are only useful with reservations. CO_2 data determined from ice cores usually show an underestimate and do not reflect "short-term" CO_2 fluctuations in the atmosphere.

Falsification of Premise 1 (man-made CO_2 increase) using official IPCC data

One does not necessarily have to use "independent data" to show that the increase in atmospheric CO_2 concentration is not caused by humans. Even with official information authorised and published by the IPPC, this is possible without much effort.

Summary for Policymakers

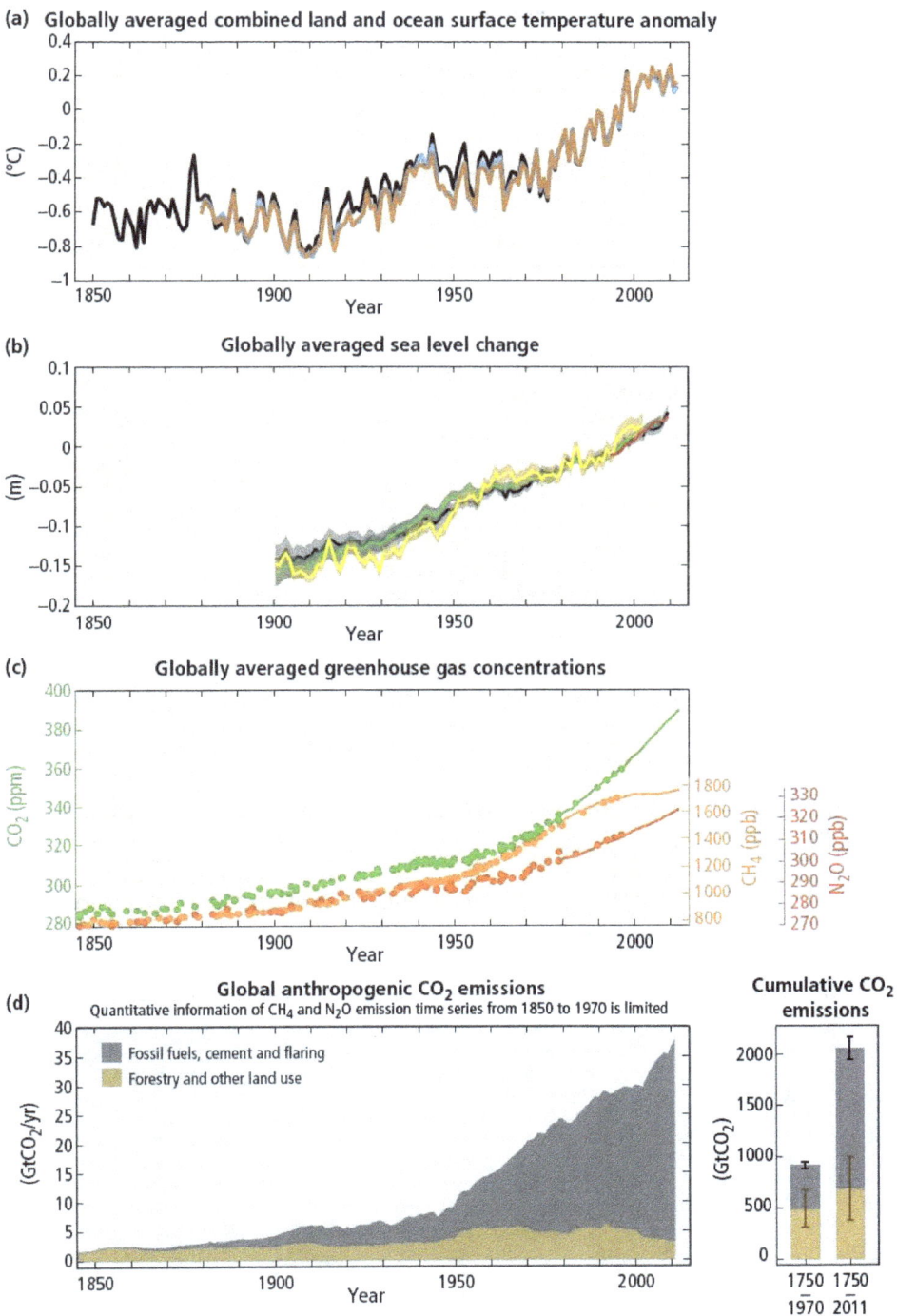

Figure SPM.1 | **The complex relationship between the observations (panels a, b, c, yellow background) and the emissions (panel d, light blue background) is addressed in Section 1.2 and Topic 1.** Observations and other indicators of a changing global climate system. Observations: **(a)** Annually and globally averaged combined land and ocean surface temperature anomalies relative to the average over the period 1986 to 2005. Colours indicate different data sets. **(b)** Annually and globally averaged sea level change relative to the average over the period 1986 to 2005 in the longest-running dataset. Colours indicate different data sets. All datasets are aligned to have the same value in 1993, the first year of satellite altimetry data (red). Where assessed, uncertainties are indicated by coloured shading. **(c)** Atmospheric concentrations of the greenhouse gases carbon dioxide (CO_2, green), methane (CH_4, orange) and nitrous oxide (N_2O, red) determined from ice core data (dots) and from direct atmospheric measurements (lines). Indicators: **(d)** Global anthropogenic CO_2 emissions from forestry and other land use as well as from burning of fossil fuel, cement production and flaring. Cumulative emissions of CO_2 from these sources and their uncertainties are shown as bars and whiskers, respectively, on the right hand side. The global effects of the accumulation of CH_4 and N_2O emissions are shown in panel c. Greenhouse gas emission data from 1970 to 2010 are shown in Figure SPM.2. {Figures 1.1, 1.3, 1.5}

The previous page (Figure 13) is taken from the IPCC's "Climate Change 2014 Report" (https://www.ipcc.ch/site/assets/uploads/2018/02/AR5_SYR_FINAL_SPM.pdf, page 3). This is the "Summary for Policymakers". This is probably why they were a little careless.

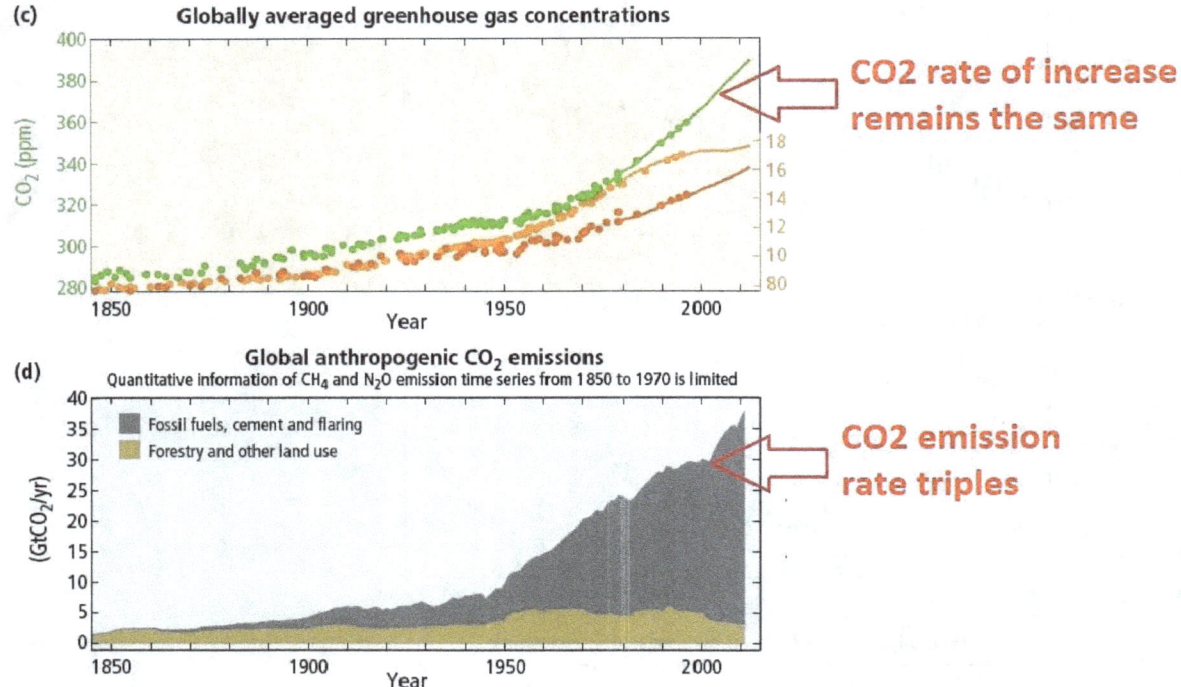

Figure 14: Something can't be right here, Source: Climate Change 2014 Report (IPCC)

Prof. Dr. Murry Salby noticed that around the year 2002, CO_2 emissions do not really match the CO_2 increase in the atmosphere (Figure 14).

In analysing the data, he notes that from the 1990s until about 2002, human CO_2 emissions from fossil fuel combustion increased at about the same rate.

After 2002, this rate increased to about three times (a consequence of the entry of China and India into heavy industry, Figure 15).

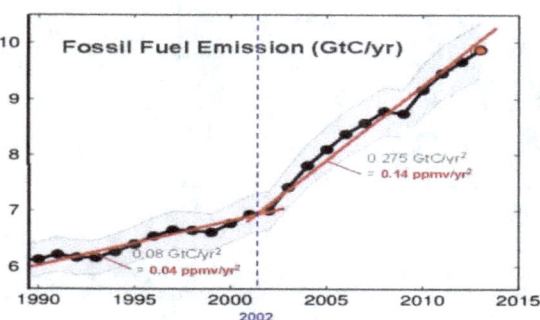

Figure 15: Source: Lecture Murry Salby London 2015 https://www.youtube.com/watch?v=jZ0R1MCkSOU

If the premise is fulfilled that the increase in CO_2 concentration in the atmosphere is only caused by humans burning fossil fuels, it would be expected that after 2002 the CO_2 concentration in the atmosphere would increase significantly faster than before.

In fact, however, something quite different is observed. In the atmospheric CO_2 concentration records, measured in Hawaii, no stronger increase can be observed after 2002. Despite the tripling of human CO_2 emissions, the increase in atmospheric CO_2 concentration remained unchanged after 2002 (Figure 16).

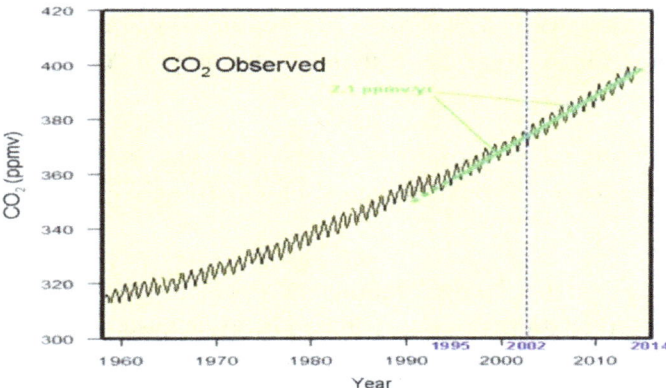

Figure 16: Lecture Murry Salby London 2015 https://www.youtube.com/watch?v=jZ0R1MCkSOU

This observation is in direct contradiction to the most important premise, on which the hypothesis of man-made global warming is based. Murry Salby was able to show that CO_2 emissions from fossil fuel combustion have no measurable influence on the increase in CO_2 concentration in the atmosphere (Figure 17).

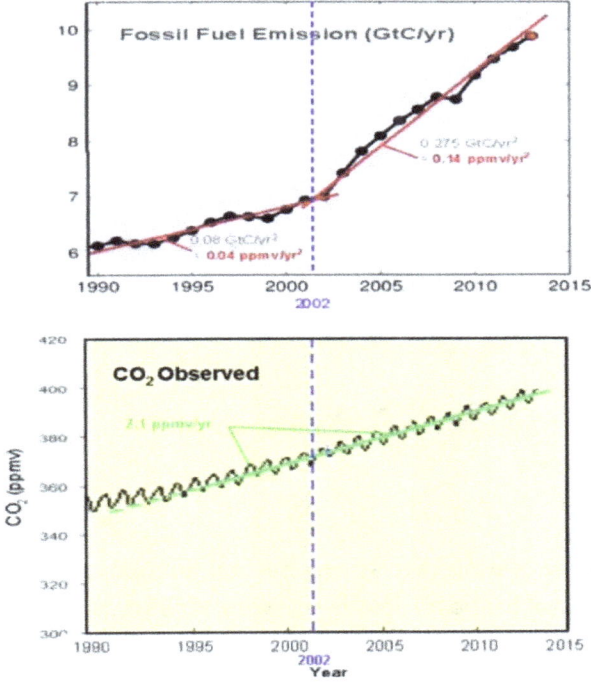

Figure 17: Lecture Murry Salby London 2015 https://www.youtube.com/watch?v=jZ0R1MCkSOU

Conversely, this means that the CO_2 concentration in the atmosphere would continue to rise even if we stopped burning fossil fuels. Or in short:

We have no measurable influence on the atmospheric CO_2 concentration.

Murry Salby has calculated the propagation of error for this observation. He comes to the following conclusion: Even if all measurement errors of the underlying measurements go in the same direction "to our disadvantage", the increase in CO_2 concentration in the atmosphere caused by humans cannot exceed one third of the observed increase.

Even assuming that the error is 50%, we would therefore not have the possibility to stop the increase of the atmospheric CO_2 concentration. **This means that all climate protection measures are ineffective because there must be another, much larger CO_2 source besides human CO_2 emissions that dominates events.**

Murry Salby has presented these results in several lectures, some of which can also be found on YouTube. I have taken the graphs shown here from a YouTube video of his lecture in London on 03.17.2015.

Salby's thesis has been criticised, sometimes harshly, by the "debunkers" (including some "climate realists"). His analysis makes do with very little data that is accessible to everyone. His argumentation is very clear and simple. In principle, any of his colleagues could have done it in a morning (without a supercomputer).

Summary, falsification of Premise 1 (man-made CO_2 increase)

Let us once again consider the two premises of the IPCC theory, whose invalidity I want to prove:

1. The steady increase in CO_2 levels in the atmosphere observed since the beginning of the Industrial Revolution is caused by humans burning fossil fuels.
2. There is an atmospheric greenhouse effect. The increase in CO_2 concentration in the atmosphere intensifies the atmospheric greenhouse effect and thus leads to a (dangerous) warming of the earth.

Premise 1 is demonstrably invalid

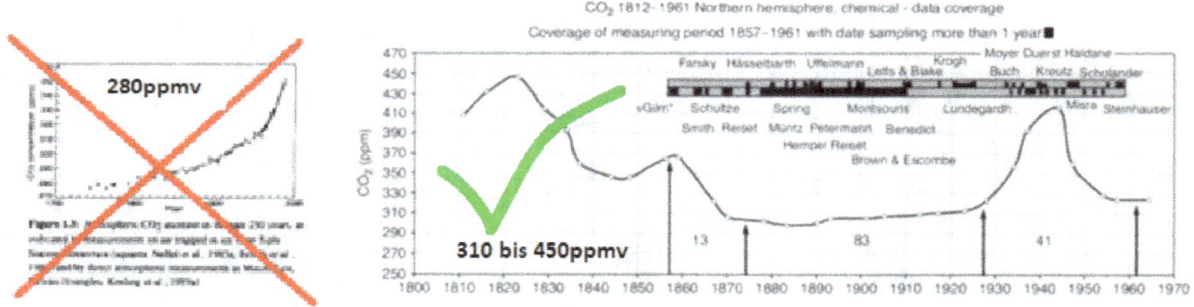

Figure 18

The CO_2 content of the atmosphere is subject to strong natural fluctuations. Even in pre-industrial times, CO_2 concentrations measured in the atmosphere, were at least as high as the CO_2 concentration of the present.

The 280ppmv of atmospheric CO_2 given by the IPCC for pre-industrial times are based on inappropriate methods and fudged measurements.

Against the background of very large natural CO_2 sources, human CO_2 emissions are so low that climate protection measures cannot have a measurable impact on atmospheric CO_2 concentrations.

The issue of climate protection is thus settled. It is not in our power to stop the increase of the atmospheric CO_2 concentration by saving CO_2.

As announced at the beginning of the text, it has been clearly shown that the "Man-made Global Warming Theory" is wrong, without having to resort to complicated arguments. Readers who simply wanted to know whether CO_2-saving climate protection measures make sense or not can stop reading at this point with a clear conscience.

For those looking for a deeper understanding of the subject, I will now go into more detail. Where it makes sense to me, I will refresh simple school knowledge and also discuss simple calculations in detail.

In advance, a few basics to refresh school knowledge

In the following, we will often speak of **energy.** Because this term is nowadays heavily used by esotericists, "psychologists" and coaches of all kinds, I would like to point out that this term has exclusively this meaning in the following: **The energy contained in a system describes the ability of that system to do work or generate heat.** Details on this can be found in Wikipedia or in textbooks.

Phase: In chemistry and physics, a phase is a spatial area in which the material properties do not change abruptly. The interface between two phases is called the phase boundary. At the phase boundary, the material properties change abruptly. This can be illustrated by a glass partially filled with Coca Cola (Figure 19).

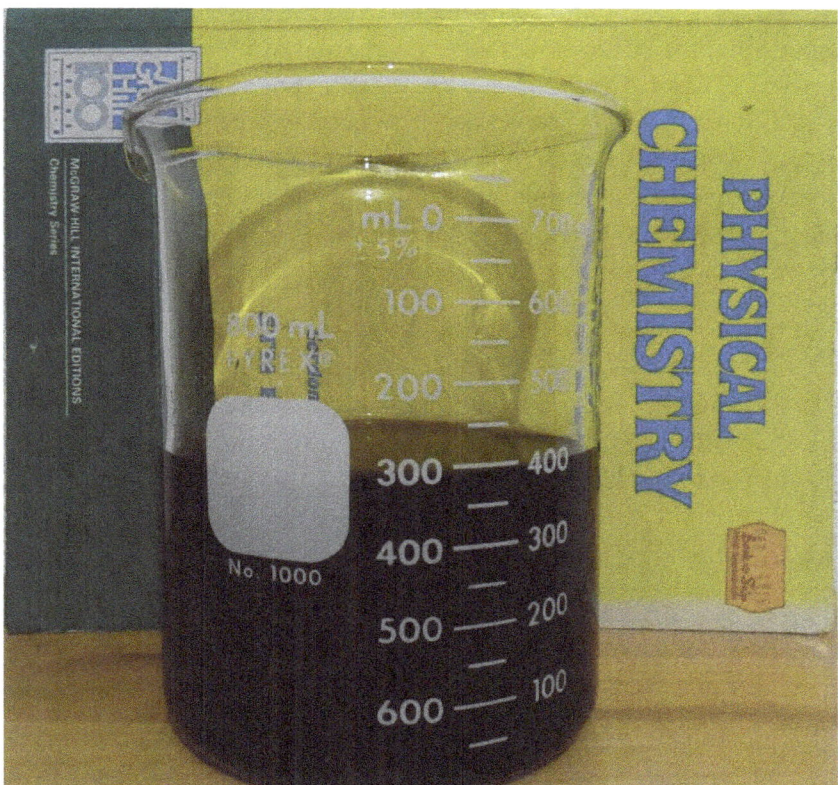

Figure 19: Half Coke Glass

The glass contains a gas phase (air) and an aqueous phase (Coca Cola). Within the respective phases, the material properties do not change. At the boundary between the two phases, there is an abrupt change in the material properties. The colour changes from colourless to dark brown. The density changes from approx. 1.2kg/m^3 in the air to approx. 1000kg/m^3 in the cola. The sugar content of the air is 0 g/litre. In Coca Cola, the sugar content is approx. 100g/litre.

How do you imagine gases, liquids and solids?
Because we will be talking about the atmosphere and the oceans in the following, we will quickly explain how to imagine gases, liquids and solids in a simple particle model.

Gases consist of very small particles, mostly small molecules. Air consists mainly of nitrogen (N_2 molecule) and oxygen (O_2 molecule). The size of these molecules is measured in the unit picometers (abbreviation: pm). A picometer has a length of 0.000 000 000 001m or 10^{-12}m. Nitrogen and oxygen molecules have a length of just over 100pm. CO_2 is a rod-shaped molecule and is about twice as long as oxygen or nitrogen.

Only very weak attractive forces act between gas particles. These particles are therefore free to move. They move very quickly. In the space available to them, they move in a criss-cross pattern. In the process, they collide with other gas particles or, if they are enclosed in a container, with the container wall. In collisions, they transfer part of their kinetic energy to other gas particles or the container wall. If you increase the temperature of a gas, the gas particles move faster. This increases the distance between the gas particles and the gas expands. If the gas is enclosed in a solid container (i.e. it cannot expand), the pressure in the container increases because the gas particles then hit the container wall at a higher speed and rate. The energy supplied to a gas during heating is stored in the kinetic energy of the gas particles. I.e. the gas particles become faster when heated.

Liquids also consist of very small particles. Liquids differ from gases in that the liquid particles attract each other more strongly. This means that the particles no longer move freely around in space. Their mobility is greatly restricted compared to gases. They remain close together and form a visible surface. As with gases, liquid particles react to heating by making their components move faster. If you heat a liquid sufficiently, the liquid particles become so fast that they can overcome the mutual forces of attraction. They then break through the surface of the liquid and move freely in space like gas particles. This process is called evaporation. When they cool down, they slow down again and return back into the liquid or form small drops. This process is called condensation.

Solids differ from gases and liquids primarily in that the particles of which they are composed attract each other very strongly. Due to the strong mutual attraction, the particles are anchored in their place in the solid. They only oscillate a little back and forth around this place. If you increase the temperature of the solid, the oscillations become larger until the particles can break away from their place. The solid starts to melt and becomes a liquid.

Under this link you can find an animation that illustrates the above well:
https://www.youtube.com/watch?v=_GzuKdPkdDQ

CO$_2$ Solution Equilibrium between Atmosphere and Oceans

Purely instinctively it is hard to believe that human CO_2 emissions should have no measurable influence on the concentration of CO_2 in the atmosphere. However, this mystery will be cleared up after a brief introduction to the solubility properties of CO_2 in the oceans. What exactly happens to CO_2 in the water of the oceans is still the subject of research and is not fully understood. However, I assume that what follows can be considered fairly well established.

CO_2 dissolves very well in the water of the oceans. It is assumed that 50 times as much CO_2 is dissolved in the oceans as is contained as a gas in the atmosphere. You can see from the nice round number 50 that the exact factor is not known. This figure is given in the CO_2 article in Wikipedia, also appears again and again in the literature, and seems to be a fairly good estimate. The exact factor doesn't really matter. The important thing is that the oceans contain much, much more CO_2 than the atmosphere. The CO_2 in the oceans is in solution equilibrium with the CO_2 in the atmosphere.

In 1802, William Henry formulated the so-called Henry (absorption) law. Details can be found in Wikipedia. Applied to our problem, Henry's law states that a gas will always distribute itself in the same ratio between a gas phase and a water phase at constant temperature and unchanging volume.

The easiest way to explain this is with a thought experiment (model):

We assume a closed container filled with nitrogen and water (Figure 20).

Figure 20:

We introduce a quantity of CO_2 into the gas phase (N2 phase) of the container (Figure 21).

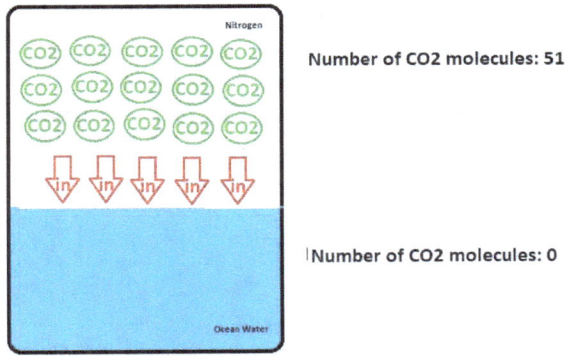

Figure 21: CO_2 is introduced into the gas phase and dissolves in the water.

Most of the CO_2 molecules then pass into the water phase due to the good solubility of CO_2 in water (Figure 22).

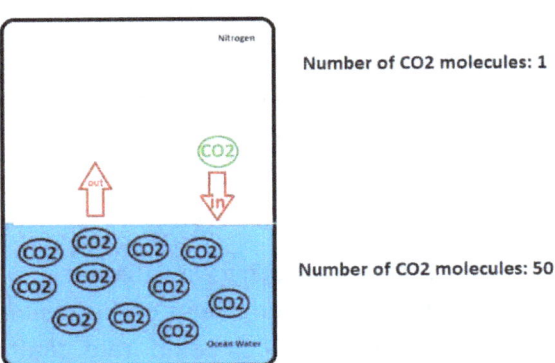

Figure 22: Dynamic equilibrium

However, CO_2 molecules can also change from the water back into the gas phase. The CO_2 molecules distribute themselves between the water phase and the gas phase. After some time, a state is reached in which, per unit of time, just as many CO_2 molecules pass from the gas phase into the water phase as vice versa. This state is called dynamic equilibrium. In dynamic equilibrium, the individual molecules switch back and forth between the two phases. However, the total amount of CO_2 molecules contained in each phase no longer changes. If we call the speed with which the CO_2 molecules change from the gas phase to the water phase V(in) and the speed of the transition from the water to the gas phase V(out), we get the ratio in which the CO_2 molecules are distributed between the two phases:

$$\frac{Number\ of\ CO2-Molecules\ in\ the\ gasphase}{Number\ of\ CO2-Molecules\ in\ the\ waterphase} = \frac{V(out)}{V(in)} \quad \text{(equation 1)}$$

Under the above premises (temperature and volume do not change), the ratio V(out)/V(in) is a constant (a number that does not change).

If we transfer this model to the atmosphere and the oceans, we get:

$$\frac{Mass\ of\ CO2\ in\ the\ Atmosphere}{Mass\ of\ CO2\ in\ the\ oceans} \approx \frac{1}{50} \quad \text{(equation 2)}$$

In other words, the oceans contain 50 times as much CO2 as the atmosphere.

This distribution equilibrium or solution equilibrium has an interesting consequence. If additional CO_2 is introduced into the atmosphere, it will be distributed in a ratio of 1/50 between the atmosphere and the oceans. I.e. of 51kg of additional CO_2 introduced into the atmosphere, 50kg will dissolve in the oceans. Only about 2% will remain in the atmosphere. In the oceans, part of the CO_2 is incorporated into lime shells by living organisms, which then form the earth's huge lime and marble deposits over time. It is therefore permanently removed from the atmosphere.

In this model (thought experiment) we assumed that the temperature does not change. This premise is certainly not fulfilled in the real world. We therefore want to generalise our model and look at how changes in the temperature of the oceans affect the distribution of CO_2 between the ocean and the atmosphere.

Temperature dependence of the solution equilibrium

The solubility of CO_2 in pure water is strongly temperature-dependent. While under normal pressure, close to freezing point, approx. 1.7 litres of CO_2 dissolve in one litre of water, at 20°C only about half of this amount dissolves (approx. 0.9 litres of CO_2 per 1 litre of water). In the oceans, CO_2 solubility is influenced not only by water temperature, but also by alkalinity (detailed information on this can be found in Wikipedia) and biological activity. However, we are not wrong, if we make the assumption that, as with pure water, it is primarily the water temperature that determines CO_2-solubility.

Applied to our model, this means: After an increase in water temperature, a new equilibrium is established. In this new equilibrium, the proportion of CO_2 in the atmosphere is higher than before. I.e. in equation 2, the ratio of CO_2 in the atmosphere to CO_2 in the oceans increases. **In warming periods, the oceans therefore release enormous amounts of CO_2** (similar to a heated soda water bottle).

Against this background, the results of Murry Salby are to be interpreted in such a way, that we are currently in a warming period of the oceans. In this warming period, the CO_2 emissions of the oceans exceed by far all human emissions. Even a dramatic increase in the human CO_2 emissions rate, can´t be detected against the background of the enormous CO_2-outgassing of the oceans.

When the oceans cool, the equilibrium shifts in the opposite direction. The oceans then absorb CO_2 from the atmosphere until a new equilibrium with a lower atmospheric CO_2 concentration is established.

This simple model sufficiently explains, on the basis of well-understood textbook chemistry and physics, why we cannot measurably influence the atmospheric CO_2 concentration.

How can it be then that the mainstream media keep warning, that we could trigger a **"Run Away Global Warmig"** through our CO_2 emissions? In other words, the exact opposite of what our model predicts.

The idea behind this doomsday scenario is the following. It is believed that we are setting in motion the following fatal mechanism:

- By burning fossil fuels, we increase the atmospheric CO_2 concentration.
- The higher CO2 concentration in the atmosphere increases the greenhouse effect.
- The increased greenhouse effect leads to a warming of the atmosphere.
- This warming causes the oceans to become warmer.
- Ocean warming releases CO_2 dissolved in the oceans.
- The CO_2 released contributes to further intensification of the greenhouse effect.
- The enhanced greenhouse effect ...

A cycle of doom with positive feedback.

Before we panic, let's look at how the oceans can warm up. I can think of four mechanisms:

1. The sun shines into the water of the oceans. The energy of the sunlight is absorbed and converted into heat.
2. Heat passes from the warm atmosphere into the cold oceans.
3. At the bottom of the oceans, water is heated by volcanic activity.
4. Frictional heat is released by water movements.

The contribution of mechanisms 3. and 4. to the warming of the oceans is very difficult to estimate. Because these processes are not directly related to "man-made global warming", I will not go into these effects.

Mechanism 1. is most certainly responsible for a very large proportion of the heat input into the oceans. Especially in the tropics, the light falls very steeply on the water surface and little light energy is lost through reflection.

Mechanism 2. would be particularly critical for a "Run Away Global Warming". If this mechanism contributes noticeably to ocean warming, the positive feedback described above could actually be a problem.

In order to estimate the extent of heat input into the oceans from the (warm) atmosphere, we carry out a rough calculation.

To keep the calculation simple, let's assume the following situation. A cubic metre of 20°C warm water and a cubic metre of 40°C warm air exchange their heat with each other. The cubic metre of air has a heat capacity of approx. 1 kJ/K. The cubic metre of water has a heat capacity of approx. 4200 kJ/K. When the air has given up its excess heat to the water, it has cooled down to approx. 20°C. During this process the heat flow from the air into the water equals approx. 20 KJ. These 20KJ warm the water warmed up by a little less than 0.005°C. How much time this heat exchange takes depends on many influences. However, one can assume that the process is slow.

If the cubic metre of water in an open pool with the dimensions 1m x 1m x 1m stands in the blazing midday sun, this amount of heat is radiated into the water in less than half a minute (solar constant: 1367w/m^2).

Result: The heat input into the oceans via heat exchange with the atmosphere is negligible compared to the heat input from solar radiation.
A "Run Away Global Warming" via this mechanism is therefore very unlikely.

This reassuring statement is also confirmed by CO_2 and temperature data obtained from ice cores.

Cause-effect relationship between the course of temperature and atmospheric CO_2 concentration

The following graph by Anthony Watts (blog, April 4, 2012, Figure 23) shows the course of temperature and CO_2 concentration over the last 800 000 years. The data for this climate data reconstruction was obtained from ice cores. I will explain how to reconstruct historical temperature data from ice cores in the next chapter.

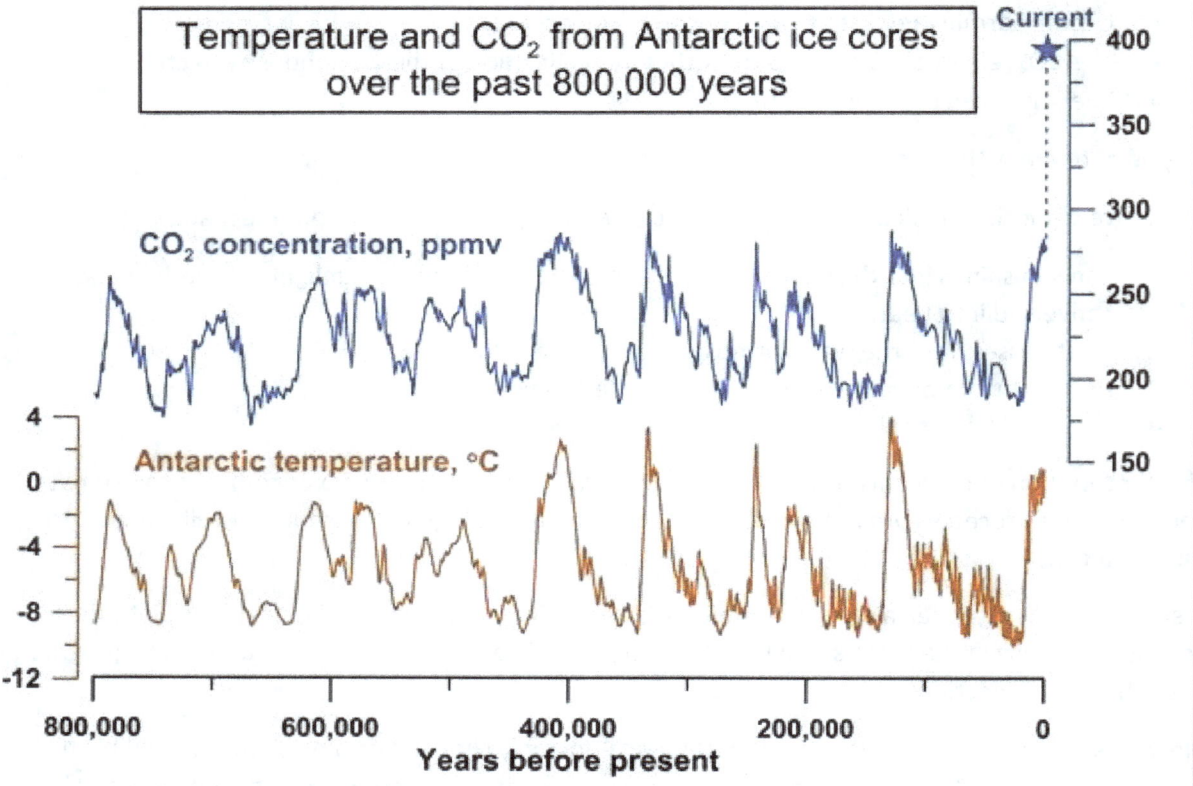

The 800,000-year record of atmospheric CO_2 from the EPICA Dome C and Vostok ice cores, and a reconstruction of local Antarctic temperature based on deuterium/hydrogen ratios in the ice. The current CO_2 concentration of 392 ppmv is shown by the blue star. (data from Lüthi et al., 2008, Nature, 453, 379-382, and Jouzel et al., 2007, Science, 317, 793-797).

Figure 23: Source: Anthony Watts

It can be seen very clearly, that temperature and CO_2 concentration always follow the same course (correlate). In addition, a "run away global warming" is nowhere to be seen. So this fits our model quite nicely.

In the graph above, however, it is also noticeable that the CO_2 concentrations determined in the ice cores are always significantly lower than the 400ppmv measured in the atmosphere today. We have already discussed this property of the ice core data.

But now it gets interesting. The climate alarmists claim that the rise in CO_2 concentration causes the rise in temperature. If this is true, the CO_2 concentration must rise somewhat earlier than the temperature, because the cause must always precede the effect in time.

If the deniers are right, it must be the other way round. I.e. first the temperature must rise, the oceans must become warmer and only then with a certain delay the CO_2 concentration must rise. Due to the enormous heat capacity of the oceans, a long lag time is to be expected.

In order to assess this, one needs data with a somewhat better temporal resolution. To illustrate this, I use a graph of ice core data (Figure 24, source: Piers Corbyn, his website weahtheraction.com is highly recommended. He is the brother of former Labour leader Jeremy Corbyn). **You can see very clearly here that, first the temperature rises and then, with a clear delay, the CO_2 concentration follows. This means that the increase in CO_2 concentration cannot possibly have been the cause of the temperature increase, which took place about 800 years earlier.**
The "man-made global warming theory" is thus refuted.

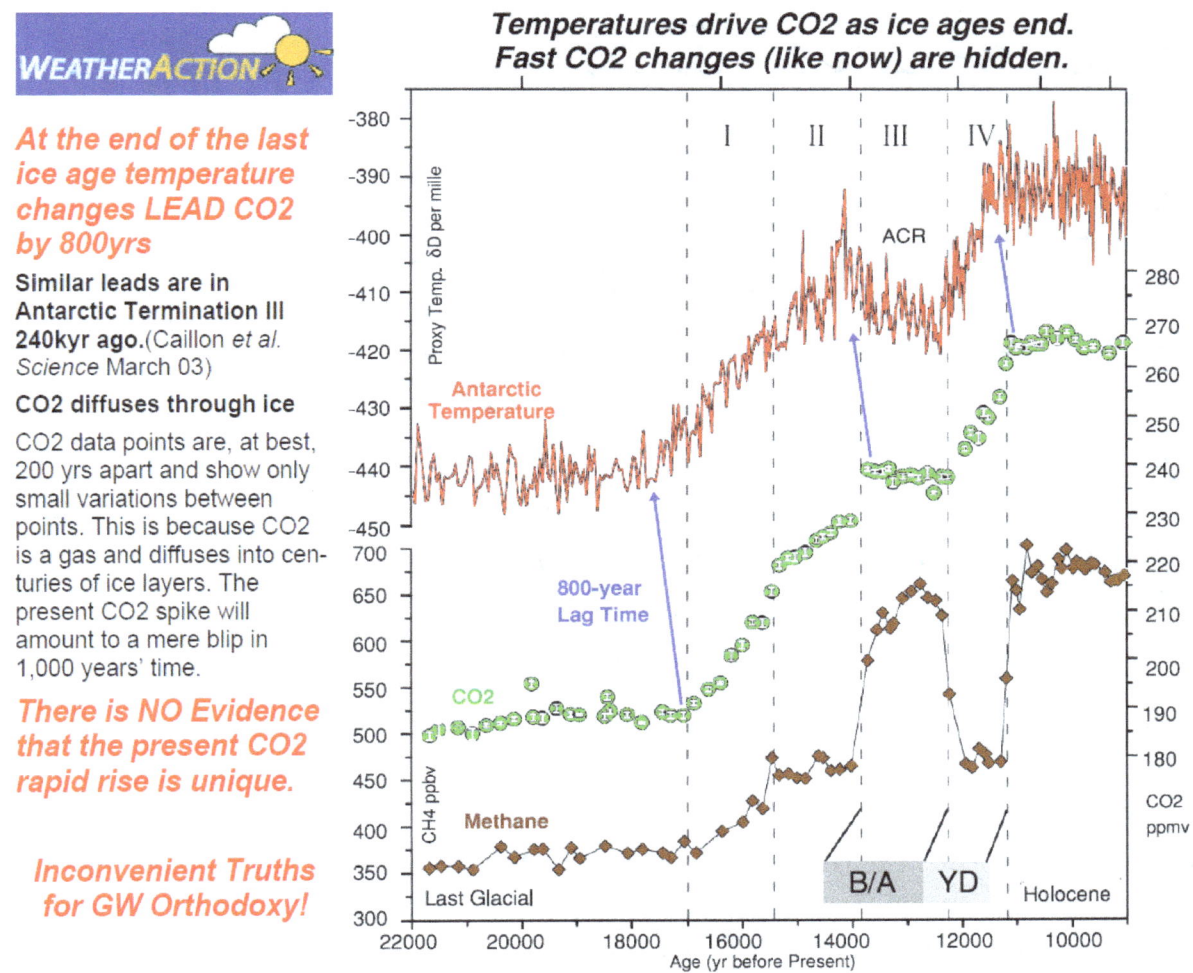

Figure 24: Source: Piers Corbyn, http://weatheraction.com/

If this pattern, which can be observed in the ice data, has a certain regularity, the CO_2 increase of our present time should also be the result of a climate warming that must have taken place about 800 years ago.

The IPCC Report 1990 provides information about the warm period (Figure 25) that is presumably responsible for today's CO2 increase. It is the much discussed Medieval Warm Period, which, if the IPPC Report 1990 is to be believed, was even warmer than our present warm period.

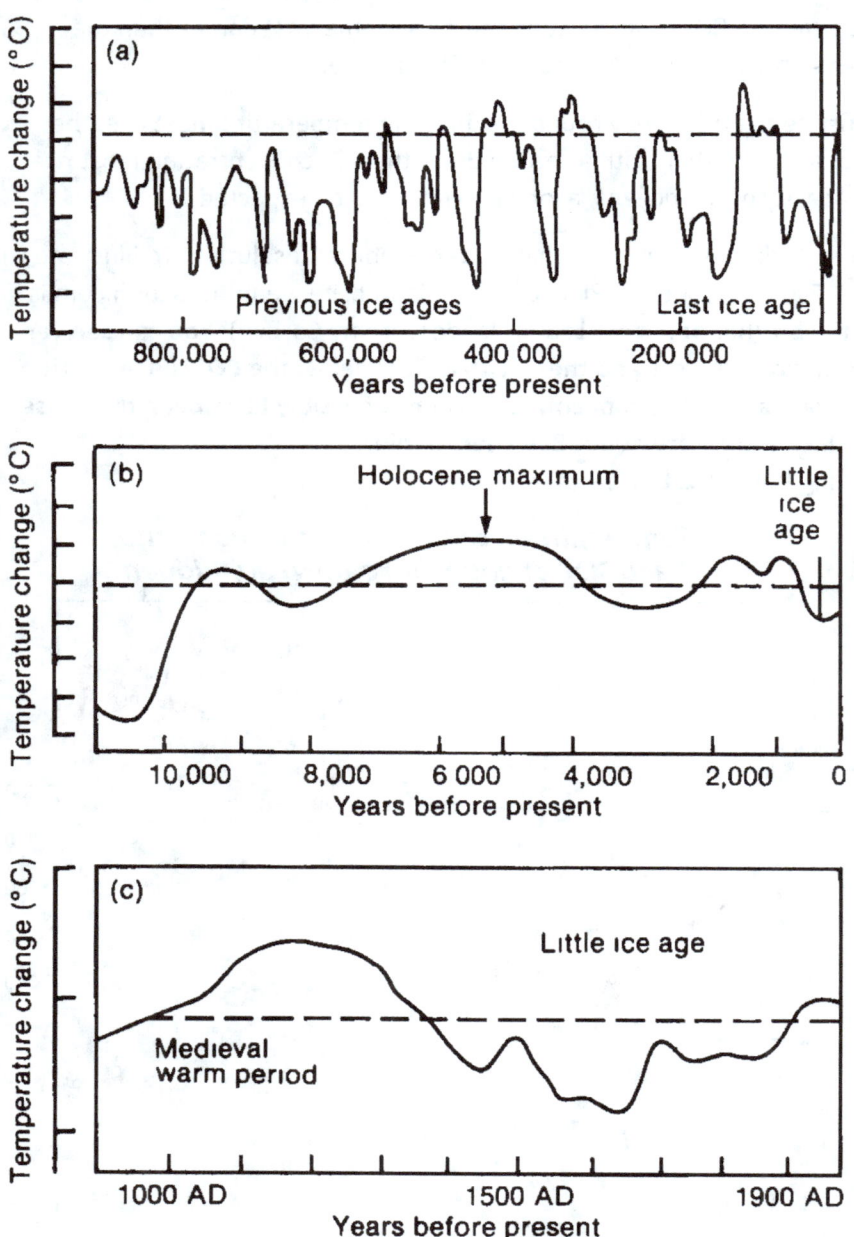

Figure 7.1: Schematic diagrams of global temperature variations since the Pleistocene on three time scales (a) the last million years (b) the last ten thousand years and (c) the last thousand years The dotted line nominally represents conditions near the beginning of the twentieth century

Figure 25: IPPC Report 1990 page 202, Medieval Warm Period, https://www.ipcc.ch/site/assets/uploads/2018/03/ipcc_far_wg_I_full_report.pdf

The astonishing result of Murry Salby's calculation thus fits easily into the climate data determined from the ice cores.

However, it still needs to be clarified how the approx. 800-year delay between warming and CO_2 increase comes about. I would like to present an attempt by Piers Corbyn to explain this. If Piers is right, ocean currents are the solution.

During the Medieval Warm Period, glaciers melted much faster than before. As already mentioned, cold water can dissolve very large amounts of CO_2. As a result, the melt water from the glaciers carries large amounts of CO_2 into the Nordic Seas. There, the meltwater sinks and flows southwards deep below the sea surface (Figure 26).

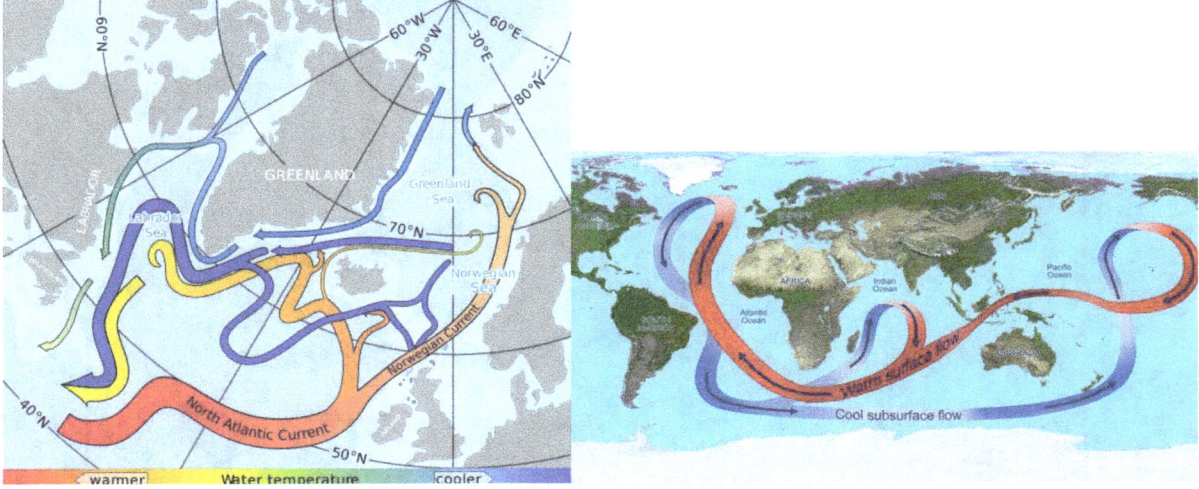

Figure 26: Source: Piers Corbyn, http://weatheraction.com/

In the depths of the ocean, it cannot release any CO_2 due to high pressure and low temperature. Only when it comes to the surface again about 800 years later in the Indian Ocean does it warm up again and release excess CO_2. And that is what we are measuring today.

Short summary of the results

Let's look again at the two most important premises of the hypothesis of man-made global warming.
1. The steady increase in CO_2 levels in the atmosphere observed since the beginning of the industrial revolution is caused by humans burning fossil fuels.
2. There is an atmospheric greenhouse effect. The increase in CO_2 concentration in the atmosphere intensifies the atmospheric greenhouse effect and thus leads to a (dangerous) warming of the earth.

We now have thoroughly checked the first premise. The result is clear.

- Human CO_2 emissions have no measurable influence on the increase in CO_2 concentration in the Earth's atmosphere.
- Temperature changes (especially of the oceans) determine the course of atmospheric CO_2 concentration (not vice versa).
- Solid measurements over the last 200 years show that the data presented by the IPCC on the course of atmospheric CO_2 concentration is wrong and that already at the beginning of industrialisation values were measured that even exceeded today's values. These high CO_2 levels cannot have been caused by industrialisation, which was still in its infancy at the time.
- The reconstruction of atmospheric CO_2 concentrations via the stomatal density of fossil plants also indicates that the "pre-industrial" value of below 280ppmV must be wrong.

We have shown that the first premise is not fulfilled and that the "man-made global warming hypothesis" must be false beyond all doubt.

The following graph, by Tony Heller, illustrates very clearly how political decisions affect the atmospheric CO_2 concentration.

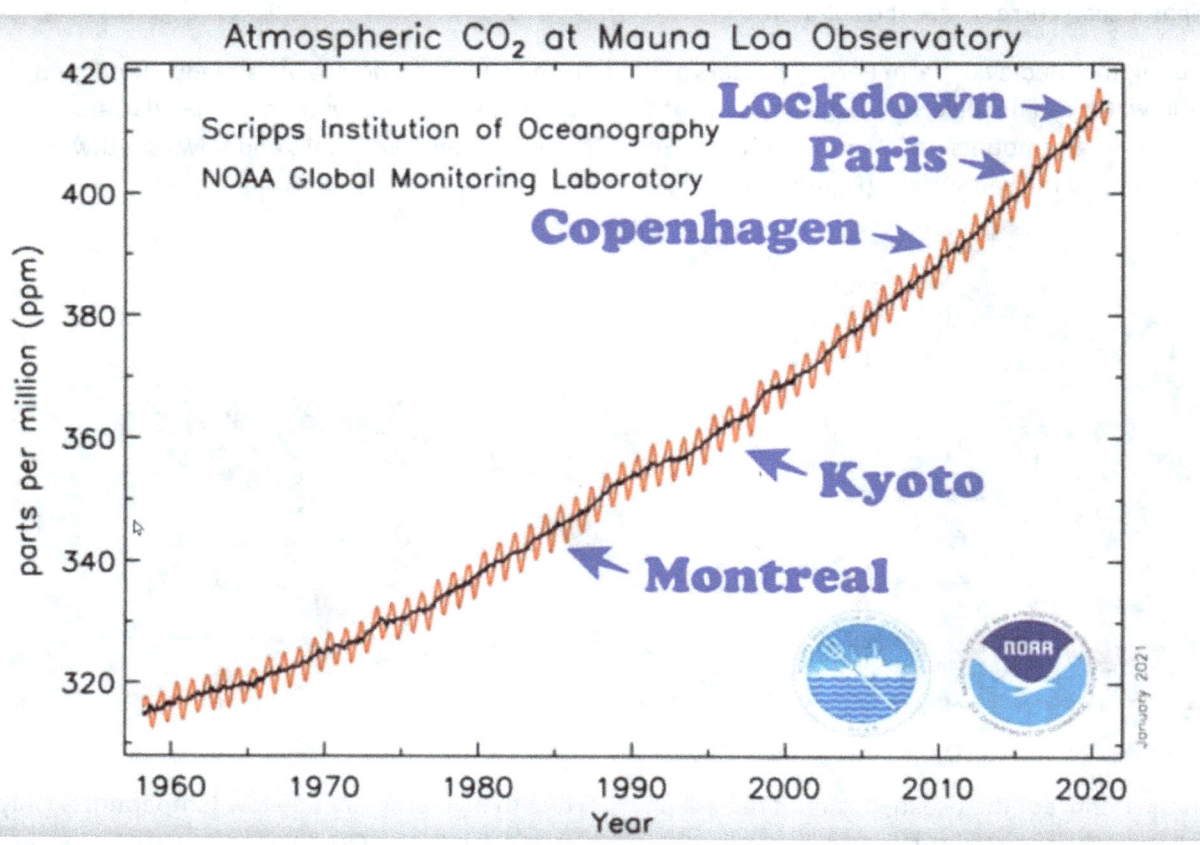

Figure 27: Source: Tony Heller https://www.youtube.com/watch?v=OCTwukaXDgw&feature=push-u-sub&attr_tag=C6hi4B-VdM0IeKbl%3A6

Reconstruction of historical atmospheric temperatures via isotope ratios

Before I review the second pillar of the man-made global warming hypothesis, the atmospheric greenhouse effect, I will briefly discuss the reconstruction of historical climate data.

So-called **proxies** are used to reconstruct climate data. These are data that are determined from historical deposits and allow statements about climate factors at the time of their deposition.

Isotope ratios in glacial ice or marine sediments are usually used as proxies for historical atmospheric temperatures.

Normal water (H_2O) contains a small proportion of water molecules that are heavier than most other water molecules. This is because these water molecules contain hydrogen atoms with an additional neutron. Or that they contain oxygen atoms with two extra neutrons. These additional neutrons have practically no influence on the chemical properties of the water. The heavier water molecules differ from normal water molecules in that they evaporate somewhat more slowly and condense somewhat more quickly than the light water molecules.

Hydrogen contains about 0.02% of the heavier version (if you don't want to embarrass yourself, say isotope instead of version).

In the case of oxygen, the proportion of the heavier isotope is approx. 0.2%.

If you want to look at literature on this topic, you should know the following denotations.
Normal hydrogen is written as ^1H. The superscript 1 indicates that this hydrogen atom is one atomic mass unit heavy. The heavy version of hydrogen is written as ^2H. The superscript 2 indicates that this atom is two atomic mass units heavy. Because ^2H is often used, it is abbrevated to D and called deuterium.
The "normal" version of oxygen weighs 16 atomic mass units (unit: u). The formula for this type of oxygen is therefore ^{16}O. The heavy version is 2u heavier. Accordingly, this is written ^{18}O.

In the water cycle (which we still know from geography lessons), water evaporates from the oceans. It is then transported over land by air currents, where it comes down again as precipitation and flows back into the sea. In the Polar Regions, it does not flow back into the sea as a river but as glaciers.

What happens to the heavier water in the water cycle? Because the heavy water evaporates a little more slowly than the normal water, the water vapour that rises from the sea contains slightly less heavy water than the water that remains in the sea. On the way to land or over land, some of the water vapour condenses again and again, forms clouds and falls to earth as precipitation (and flows back into the sea). Because heavy water condenses faster than normal water, the water vapour loses proportionally more heavy water than light water during each of these precipitation events. When the water vapour is transported by cold air masses, the heavy water is more thoroughly removed from the vapour than in warm air masses. This process, the separation of a component of the vapour by evaporation and condensation, is called fractionation (as in distilling schnapps).

In periods of cold climate, a higher proportion of heavy water is thus lost on its way from the oceans to the glaciers than in periods of warm climate. In the glacier ice deposited during cold periods, the proportion of heavy water is therefore smaller than in the ice deposited during warm periods.

Mass spectroscopy can be used to determine the isotope ratios of $^2H/^1H$ or $^{18}O/^{16}O$ in ice samples. After a suitable calibration, these ratios can be converted into temperatures. Often, instead of the temperatures, only the $^2H/^1H$ or $^{18}O/^{16}O$ isotope ratios of the investigated samples are given. This is then written as "proxy Temp. δD per mile" or "proxy Temp. $δ^{18}O$ per mile" (see Figure 21).

$^{18}O/^{16}O$ isotope ratio in marine sediments as a proxy for Earth's degree of glaciation.
From what has been said before, it is clear that especially during cold periods, heavy water rains/snows more completely from the water vapour contained in the air. The heavy water flows back into the sea, while the light water falls as snow on the glaciers. During ice ages, very large amounts of water evaporated from the oceans are deposited on glaciers. As a result, the amount of water in the oceans becomes much smaller and the sea level falls (by over 100m in the last ice age). During ice ages, the proportion of heavy water in the oceans therefore increases. Marine organisms that form calcareous shells therefore incorporate more heavy oxygen (from heavy water) into their calcareous shells during ice ages. After the death of these organisms, the calcareous shells form deposits (sediments). In drilling cores of these deposits one can then determine the oxygen isotope ratio and the age of the deposit. A high $^{18}O/^{16}O$ ratio in oceanic sediments indicates that the sediments studied were deposited during a time when a large part of the (light) water was bound in glaciers. The degrees of glaciation of the Earth determined from the oceanic sediments fit very well in time with the δD- and $δ^{18}O$-temperature proxies determined from the glacier drilling cores.

Falsification of the greenhouse effect

The greenhouse effect is the primordial sacrament of modern climate science. It is preached from kindergarten to the nursing home. The simple name of this effect and its ubiquity in culture and mainstream media leave no doubt, that we are dealing here with solid and fully understood natural science.

But how does this effect work? How is this effect derived from basic physical principles? How can the effect be measured?

These simple questions should be clear and easy to answer.

I must disappoint the reader. The atmospheric greenhouse effect is and remains a mystery.

The **German Meteorological Society** (1995) opens a statement in which, referring to the IPCC's argumentation, it assures the public that the greenhouse effect really exists with the sentence: "**It is indisputable that the anthropogenic greenhouse effect has not yet been proven beyond doubt**".
https://idw-online.de/de/news14359

This is astonishing. A branch of science sponsored with many billions of taxpayers' money is not able to trace its basic premise back to fundamental physical principles or even to measure them. With this introduction, the professors are keeping the back door open. They want to claim the academic right "to be allowed to err" if the fraud is uncovered.

In climate science there are many explanations for the atmospheric greenhouse effect, some of which are contradictory. A good overview of these explanations can be found in **"Falsifcation Of The Atmospheric CO2 Greenhouse Effects Within The Frame Of Physics" Version 4.0 (January 6, 2009) Prof. Dr. Gerhard Gerlich, Dr. Ralf D. Tscheuschner.**
https://de.scribd.com/document/337186171/Falsification-of-the-Atmospheric-CO2-Greenhouse-Effects-Within-the-Frame-of-Physics

The fact that the atmospheric greenhouse effect is nowhere "properly" explained and that there is no measurement specification for the effect does not stop modern climate science from telling us that the atmospheric greenhouse effect is warming our world by 33°C.

If we want to know what the atmospheric greenhouse effect is all about, we have no choice but to look at how the IPCC arrives at the much-cited 33°C greenhouse effect.

Before we laymen are able to understand this "top performance" of modern climate science, we need to discuss a few basic concepts of thermodynamics (absolute temperature, laws of thermodynamics, heat conduction, Stefan-Boltzmann law, ...).

For readers who are familiar with these things, I recommend skipping the following chapter and continuing reading with chapter "Stefan-Boltzmann Law".

A few basic concepts of heat theory (thermodynamics)

Absolute temperature
Let's start with the absolute temperature. At the beginning of the systematic investigation of the properties of gases, Robert Boyl (1665) investigated the heat dependence of the volume of gases at constant pressure.
In practice, you can do this by enclosing a volume of gas (which can be quite normal air) in a syringe (with a very smooth-running piston). Then you change the temperature of the syringe and the gas in it. The volume of the gas can be read off the scale of the syringe. In this way, you can measure the volume of gas as a function of its temperature. Repeat this experiment with different amounts of gas

and plot the experimental results in a coordinate system (Figure 28, temperature on the horizontal axis, volume on the vertical axis). You will find, that for each sample volume the measured values lie on a straight line. The interesting thing about these straight lines is, that they all intersect at one point on the (horizontal) temperature axis. The point of intersection is at approx. T = -273°C.

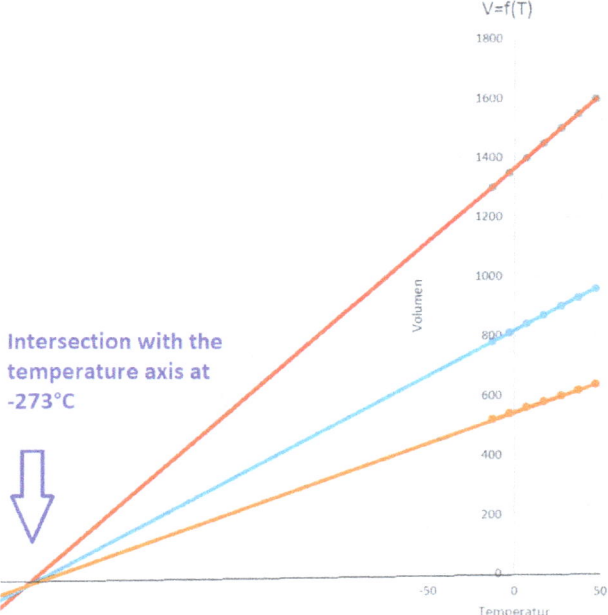

Figure 28: Gas volume as a function of temperature

This result becomes understandable if we consider again how we described a gas before. At high temperatures, the gas particles fly around very quickly and take up a lot of space. The particles have a high kinetic energy at high temperatures. When the gas is cooled, the gas particles become slower (lose kinetic energy) and take up less space (gas volume becomes smaller). With further cooling, the particles eventually become so slow that they almost stop moving. The volume of the gas becomes very small (practically zero) at this temperature. Because the gas particles no longer move at this temperature, it cannot be cooled any further. This point (-273.15°C) has been defined as the zero point of the absolute temperature scale.

A capital T is used as the symbol for the absolute temperature. The unit of absolute temperature is Kelvin [K]. One degree Celsius and one degree Kelvin have the same absolute value. The conversion of Celsius temperatures into degrees Kelvin is therefore simple.

Temperature [K] = Temperature [°C] + 273.15

Examples: Convert 0°C to Kelvin: Temperature [K] = 0 [°C] + 273.15 = 273.15K

-100°C corresponds to 173.15K

The temperature of a substance is a measure of the kinetic energy of the particles of which the substance consists. At the temperature T = 0K (-273.15°C), the particles practically no longer move.

Main laws of thermodynamics (heat theory)

Now the most important things about the main laws of thermodynamics. As the name "main laws" suggests, these theorems form the most important basis of thermodynamics. The development of these theories in the 19th century went hand in hand with industrialisation. It was this fundamental understanding of energy, heat and its transformation into useful work that enabled the development of our modern world. The main theorems of thermodynamics are scientific theories that are only valid until they are disproved. However, I would still put them in the category of "eternal truths". The laws of thermodynamics are probably the most robust theories the world has ever seen.

0st law: Describes thermal equilibrium. It states that objects that are in thermal equilibrium with each other have the same temperature.

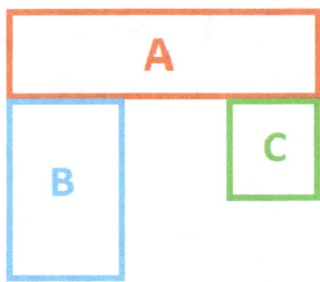

Figure 29:

I.e. in thermal equilibrium, objects A and B and A and C have the same temperature (Figure 29). Therefore, objects B and C must also have the same temperature.

1st law: Conservation of energy in a closed system. States that in a system that does not exchange energy with its surroundings, the total energy must remain the same.

2nd law: Heat cannot transfer by *itself* from a body of lower temperature to a body of higher temperature.

3rd law: Non-attainability of absolute zero (not so important in this context).

Heat conduction:
Heat always flows from objects with a higher temperature to objects with a lower temperature (2nd law). No heat is lost in the process (1st law). The heat flow comes to a standstill when all objects involved have the same temperature (0st law, thermal equilibrium).
There are various mechanisms of heat conduction that you can read about in textbooks or in Wikipedia. Of particular interest for our topic are heat conduction in gases (diffusion and convection) and heat conduction by thermal radiation.

Heat conduction by diffusion: To understand this process, we again use our "gas model". We imagine a gas enclosed in a container. If the container and the gas have the same temperature, the gas particles transfer as much kinetic energy to the wall when they collide with it as they absorb from the particles vibrating in the container wall (thermal equilibrium). If one now heats a part of the container wall, the particles of the wall vibrate more strongly at this point (have more kinetic energy). If this point of the wall is hit by gas particles, part of the now higher kinetic energy of the wall particles is transferred to the gas particles. After the impact with the hot area of the wall, the gas particles have more kinetic energy than before the impact. They can then pass on this additional kinetic energy in collisions with other gas particles. This process transports heat energy from the wall to the inside of the gas. Gases whose particles move particularly fast therefore also conduct heat particularly well (e.g. hydrogen).

Heat conduction by convection: To understand convection, we stick to the same model idea. The only difference, we make the container larger and heat a slightly larger area of the wall (Figure 30). Near the heated wall area, the mechanism described above creates a volume where the gas is warmer than in the rest of the container. In this volume, the gas particles move faster and take up more space. As a result, the gas volume near the warm wall surface contains fewer gas particles than a gas volume of the same size with a lower temperature. The warm gas volume is therefore lighter than a corresponding volume at a greater distance from the warm wall surface. The heated gas volume at the heated wall location therefore rises and is replaced by colder gas from the surroundings. This new volume is also heated and begins to rise. In this way, a so-called convection flow starts in the container.

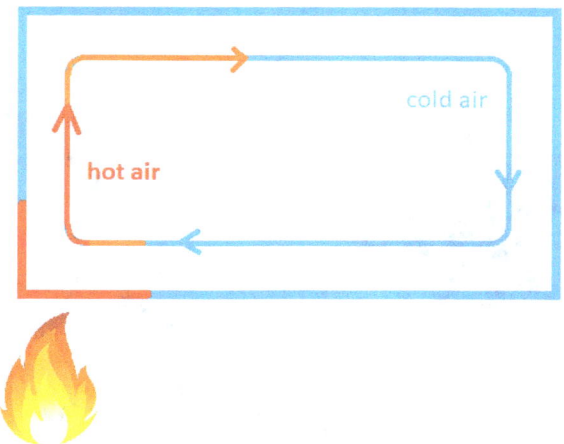

Figure 30: Convection

Such convection currents also form in the landscape when the sun heats the terrain unevenly (Figure 31).

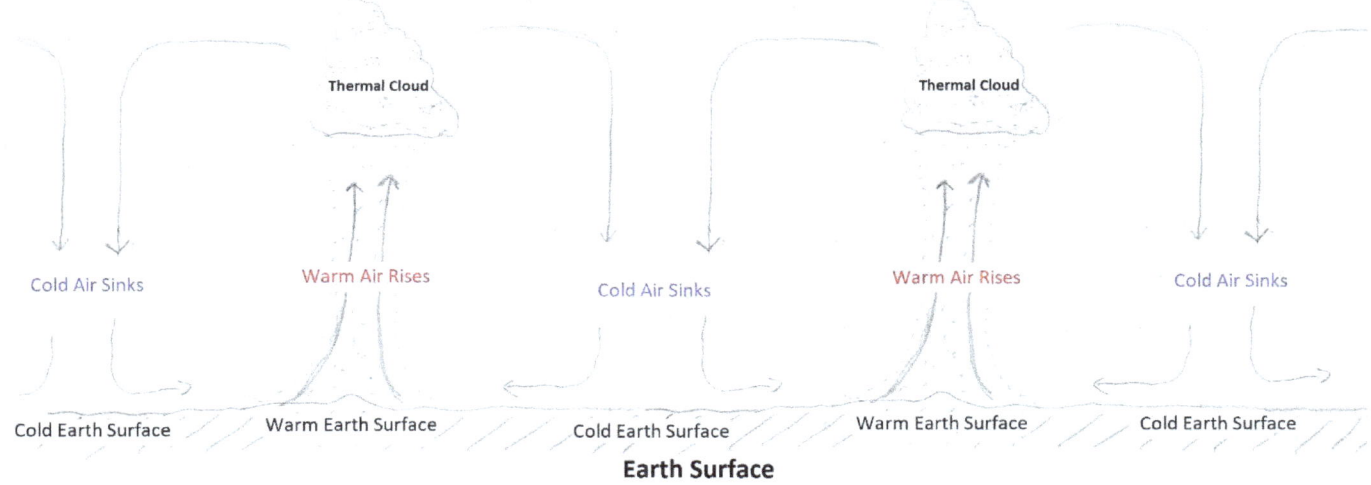

Figure 31: Convection currents in the landscape

Convection transports very large amounts of heat very quickly from the earth's surface to higher layers of the atmosphere. These convection currents are so strong that they can be used by birds and gliders. The rising part of the convection current (thermal) can reach climbing speeds of well over 100km/h (in thunderstorms). Clouds often form in the upper part of the rising convection currents (thermals) (Figure 32).

Figure 32: Thermal clouds

Heat conduction by thermal radiation: Thermal radiation is electromagnetic radiation. Details can be found in textbooks or in Wikipedia. Therefore, I will only go into this topic here as far as it is absolutely necessary for understanding heat radiation.

Electromagnetic radiation can be explained quite well using the example of a rod antenna.

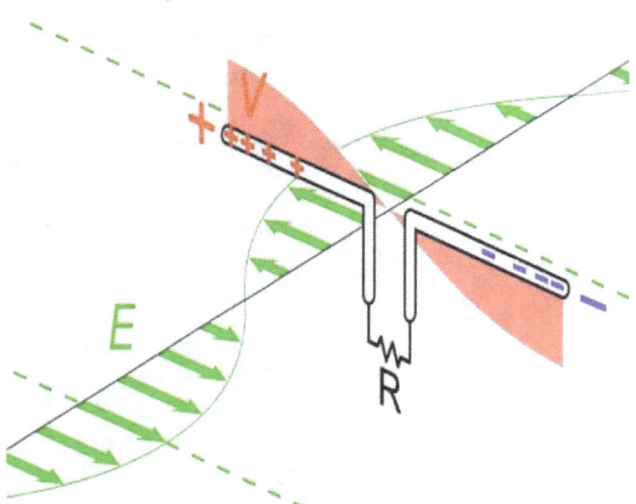

Figure 33: Antenna, source: Wikipedia (the black straight line running through the picture from bottom left to top right is supposed to be the time axis). A very clear animation can be found at this link:
https://de.wikipedia.org/wiki/Antenne#/media/Datei:Dipole_receiving_antenna_animation_6_800x394x150ms.gif

In a rod antenna, electric charges are periodically shifted back and forth. This results in a periodically changing electric field (E). The change in the E field induces a magnetic field (B) that is perpendicular to the E field (see Figure 34). The antenna therefore radiates an electromagnetic wave. The frequency of this wave depends on how fast the electric charges in the rod antenna are shifted back and forth. The energy of the wave is stored in the electric and magnetic field of the wave. Electromagnetic waves travel through space at the speed of light.

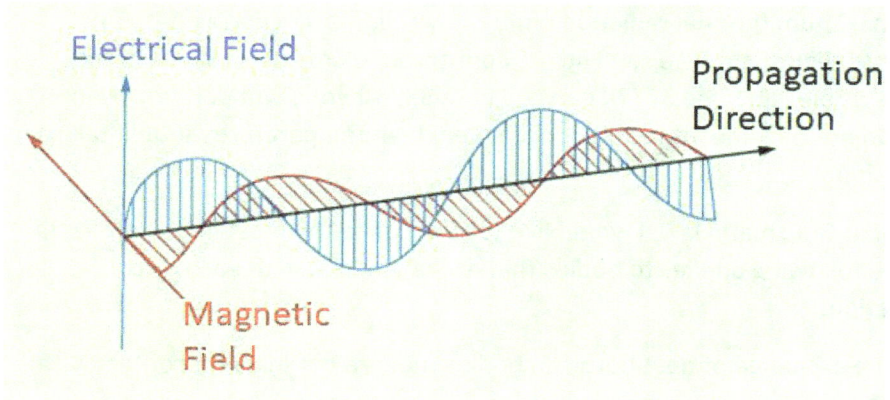

Figure 34: Electromagnetic Wave

When this electromagnetic wave hits an antenna that has similar properties to the antenna from which it was emitted, it can shift charges back and forth in this antenna and thus transfer its energy to the receiving antenna.

Solids consist of charged particles. These particles oscillate back and forth in their place. Similar to the antenna, this oscillation of charged particles generates electrical fields that change over time. Charged particles oscillating in a solid therefore emits electromagnetic radiation.

Each particle in a solid vibrates slightly differently from the particles in its environment. The solid therefore does not emit a sharp frequency like the antenna but a whole spectrum of frequencies. When the solid gets warmer, the particles vibrate faster and the spectrum shifts to higher frequencies or shorter wavelengths (see Figure 35).

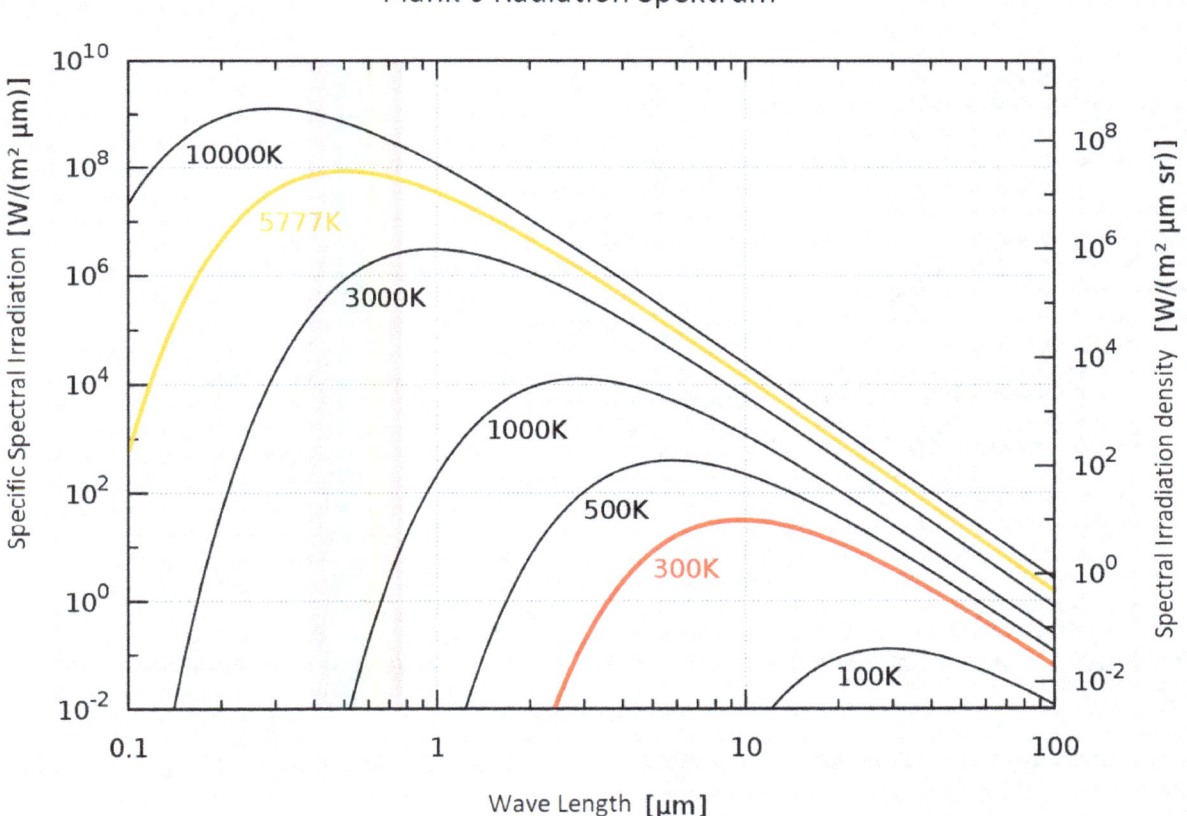

Figure 35: Spectrum of the black body, source: https://de.wikipedia.org/wiki/Schwarzer_K%C3%B6rper

At room temperature, the maximum thermal radiation emitted by solids is at a wavelength of approx. 10μm (see 300K curve, Figure 35). The sun with its approx. 6000K mainly radiates in the range of approx. 0.5μm, the visible light (see 5777K curve, Figure 35). Anyone wondering why in Figure 32 the wavelengths do not become arbitrarily short at very high temperatures should take a look at Planck's law of radiation.

Before we deal with the Stefan-Boltzmann law, I would like to summarise the most important facts about thermal radiation. The following applies to bodies that exchange heat with each other exclusively via thermal radiation:

Thermal radiation enables the exchange of heat between bodies that are not in direct contact with each other (e.g. sun and planets).

All solids warmer than absolute zero (-273.15°C or 0K) emit thermal radiation.

Heat radiation absorbed by a solid increases the heat content of the solid and thus its temperature.

The warmer a body is, the greater the amount of heat it emits as thermal radiation per unit area and time.

When a cold and a warm body exchange heat radiation with each other, more heat is transferred from the warm to the cold body per unit of time than vice versa (Figure 36).

When the temperatures of the bodies have equalised (T1 = T2), the two bodies exchange the same amount of heat with each other per unit of time. This state is called thermal equilibrium (Figure 33). **In thermal equilibrium, a body radiates as much thermal radiation per unit of time as it absorbs from its surroundings**.

The shape, colour and surface texture of the bodies do not matter. These factors only have an influence on how quickly the exchange of heat takes place and how quickly the thermal equilibrium is established.

In classical thermodynamics, time is not a parameter.

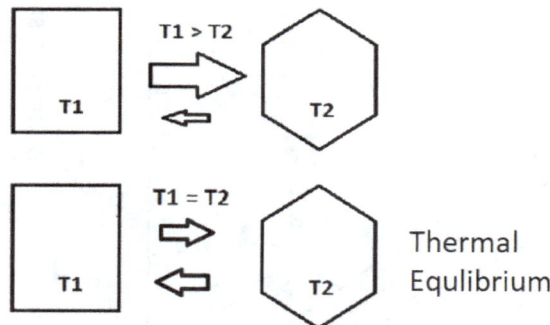

Figure 36:

Stefan Boltzmann Law

After this rather superficial excursion into thermodynamics, we can look at the Stefan-Boltzmann law.

If the temperature of a solid is known, the Stefan-Boltzmann law allows the power of the thermal radiation emitted by this solid to be calculated.

For the radiated power, I use the formula symbol P. In this case, the power (P, unit Watt [W =J/sec]) is the amount of heat (Q, unit Joule [J]) emitted per unit of time (t, unit seconds [sec]) as thermal radiation.

$P = \frac{Q}{t}$; Unit Watt [W]

We call the area over which the solid emits thermal radiation A (unit [m²]).

Let the temperature of the body be T (unit [K])

Let the temperature of the environment be T_u (unit [K])

The Stefan-Boltzmann law then reads:

P = Aσ(T^4-T^4_u)   with σ = 5.670 x 10-8 $Wm^{-2}K^{-4}$, so-called Stefan-Boltzmann constant

In the following, we want to describe the radiation behaviour of the Earth with the Stefan-Boltzmann law.
Since the Earth radiates into 0K cold space (T_u = 0K), the following equation simplifies to
P = AσT^4
Or resolved for T.

$T = \sqrt[4]{\frac{P}{A\sigma}}$

This law only applies to solids that are in thermal equilibrium. In thermal equilibrium, solids have the same temperature over their entire surface.
Moreover, it is only applicable to so-called "black bodies". These are bodies that absorb all the radiation that hits them. For bodies that do not absorb all the radiation that falls on them (so-called grey bodies), suitable correction factors must be inserted. For the Earth, a correction factor of approx. 0.7 has been determined by satellite measurements. This means that the Earth reflects about one third of the radiation falling on it from the Sun back into space. The **earth's reflectivity is called albedo (α = 0.3).**

When we calculate a planet's surface temperature via the Stefan-Boltzman-law, we assume that the planet´s surface is heated exclusively by the incoming radiation. The incoming radiation energy is absorbed by the surface, converted into heat and then radiated back into space as thermal radiation. After a heating-up phase the surface irradiates as much energy per unit of time in form of thermal radiation as it absorbs in the same unit of time from the incoming radiation, which means the surface is in thermal equilibrium with its surrounding. Therefore **we can state in thermal equilibrium**:

$$Power\ of\ the\ absorbed\ radiation = Power\ of\ the\ emitted\ thermal\ radiation$$
$$or$$
$$P_{absorbed} = P_{emitted} = (1-\alpha)A\sigma T^4$$

Incorrect application of the Stefan-Boltzmann law

In the equation P = AσT^4 the temperature is to the fourth power. As a result of that even small changes in the surface temperature have a dramatic effect on the power emitted from the surface. Thus attempts to calculate the power emitted from solids with an inhomogeneous surface-temperature-distribution by inserting an average surface-temperature into the Stefan-Boltzmann law fail dramatically.

For the same reason, it is not possible to correctly calculate an average surface temperature of a solid which is exposed to an inhomogeneous irradiation by inserting an average irradiation into the Stefan-Boltzmann law.

The easiest way to show this is with a calculation example:

A sports facility consists of two closely spaced soccer fields (with black surfaces), each with an area of 7000m². We simply calculate as if the two fields were really black, i.e. albedo (α = 0). One field is in the sun and its surface is 50°C (323K) warm. The other square is shaded and only 20°C (293K) warm.
For the field in the sun, this results in a radiation power of: 4320074W or 617W/m²
For the "shaded field": 2925173W or 418W/m²
If we consider the two squares together as one radiating unit, this results in an area-averaged radiated power of 517.5W/m² for this area of 14000m².

Now we calculate the average temperature for the total area of the two fields in two different ways.

First the correct way. We calculate the area mean temperature (T_m) based on the measured surface temperatures.

$$T_m = \frac{T_{Sunnyfield} * A_{Sunnyfield} + T_{Shadowfield} * A_{Shadowfield}}{A_{Sunnyfield} + A_{Sadowfield}}$$

$$T_m = \frac{323K * 7000m^2 + 293K * 7000m^2}{14000m^2} = 308K = 35°C$$

Then we use the Stefan-Boltzmann law to calculate the area-averaged temperature of the sports facility from the area-averaged radiated power of the two courts (P/A=517.5W/m²).

$$T_m = \sqrt[4]{\frac{P}{A\sigma}} = \sqrt[4]{\frac{517{,}5 \frac{W}{m^2}}{5{,}67 * 10^{-8} \frac{W}{m^2 K^4}}} = 309K = 36°C$$

It is noticeable that the surface temperature determined via the mean value of the radiated power deviates from the actually measured, area-averaged temperature.

Based on this small example, one can estimate what is to be thought of calculating a world average temperature based on an irradiation power averaged over the whole world.

As we will see in the following chapter, this is exactly the way IPCC-climate science determines the greenhouse effect.

Determining the greenhouse effect (according to IPCC)

Now we have the basic knowledge to understand how IPCC climate science "calculates/determines" the greenhouse effect. The basic idea behind this calculation is simple.
1. With the help of the Stefan-Boltzmann law and the known radiation power of the sun, one calculates **how warm the earth should be without an atmosphere.**
2. The real average temperature of the earth is measured via weather stations.

3. From the measured average temperature of the earth, one subtracts the temperature calculated for the earth without an atmosphere. This gives a temperature difference that is claimed to be caused by the atmospheric greenhouse effect.

$$T_{Greenhouse\,effect} = T_{global\,measured\,mean} - T_{Earth\,without\,atmosphere}$$

I will not go into detail at this point about the fact that determining a global average temperature is a very questionable matter. The necessary dense network of measuring stations simply does not exist and the existing measuring stations do not always provide reliable data. As we will see, it doesn't really matter whether the global average temperature is accurate to within a few degrees or not.

Now to the actual calculation:
In the Fourth Assessment Report Climate Change 2007 The Physical Science Basis Chapter 1: Historical overview of climate change science page 97
https://www.ipcc.ch/site/assets/uploads/2018/05/ar4_wg1_full_report-1.pdf there is, so to speak in a subordinate clause, a reference to how the IPCC determines the greenhouse effect (Figure 37). It is assumed that the Earth receives about 240W/m² of radiation from the sun. This radiation is absorbed, warms the Earth and is then re-emitted from the Earth into space as infrared radiation. With these 240W/m² one goes into the Stefan-Boltzmann law and calculates the temperature that the earth should have without the atmosphere and thus without the atmospheric greenhouse effect. The result is -19°C for an Earth without greenhouse effect/atmosphere.

> The energy that is not reflected back to space is absorbed by the Earth's surface and atmosphere. This amount is approximately 240 Watts per square metre (W m⁻²). To balance the incoming energy, the Earth itself must radiate, on average, the same amount of energy back to space. The Earth does this by emitting outgoing longwave radiation. Everything on Earth emits longwave radiation continuously. That is the heat energy one feels radiating out from a fire; the warmer an object, the more heat energy it radiates. To emit 240 W m⁻², a surface would have to have a temperature of around −19°C. This is much colder than the conditions that actually exist at the Earth's surface (the global mean surface temperature is about 14°C). Instead, the necessary −19°C is found at an altitude about 5 km above the surface.
>
> The reason the Earth's surface is this warm is the presence of greenhouse gases, which act as a partial blanket for the longwave radiation coming from the surface. This blanketing is known as the natural greenhouse effect. The most important greenhouse gases are water vapour and carbon dioxide. The two most abundant constituents of the atmosphere – nitrogen and oxygen – have no such effect. Clouds, on the other hand, do exert a blanketing effect similar to that of the greenhouse gases; however, this effect is offset by their reflectivity, such that on average, clouds tend to have a cooling effect on climate (although locally one can feel the warming effect: cloudy nights tend to remain warmer than clear nights because the clouds radiate longwave energy back down to the surface). Human activities intensify the blanketing effect through the release of greenhouse gases. For instance, the amount of carbon dioxide in the atmosphere has increased by about 35% in the industrial era, and this increase is known to be due to human activities, primarily the combustion of fossil fuels and removal of forests. Thus, humankind has dramatically altered the chemical composition of the global atmosphere with substantial implications for climate.

Figure 37, Source: IPCC Fourth Assessment Report Climate Change 2007 The Physical Science Basis Chapter 1: Historical overview of climate change science page 97 https://www.ipcc.ch/report/ar4/wg1/historical-overview-of-climate-change-science-2/

$$T = \sqrt[4]{\frac{P}{A\sigma}} \qquad \text{with P/A = 240W/m}^2 \text{ and } \sigma = 5.670 \times 10^{-8} \text{ Wm}^{-2}\text{K}^{-4}$$

$$T = \sqrt[4]{\frac{240\frac{W}{m^2}}{5{,}67*10^{-8}\frac{W}{m^2K^4}}} = 255K = -18°C \text{ (In text -19°C. Doesn't matter. It's not that exact here)}$$

This raises the question of how the IPCC arrives at the 240W/m² of irradiation. Calculating with such an average value is certainly not entirely correct, as we saw in the calculation example with the soccer fields. But regardless of this error, let's take a look at how this average value is arrived at.

The power radiated by the sun onto the earth is quite constant. It is the so-called solar constant S_0 = 1367 W/m². This value was determined by satellite measurements. For the following calculations, it is assumed that the earth is hit by solar radiation that passes vertically through a circular area with the radius of the earth's sphere. 1367 W of radiation pass through each square metre of this area.

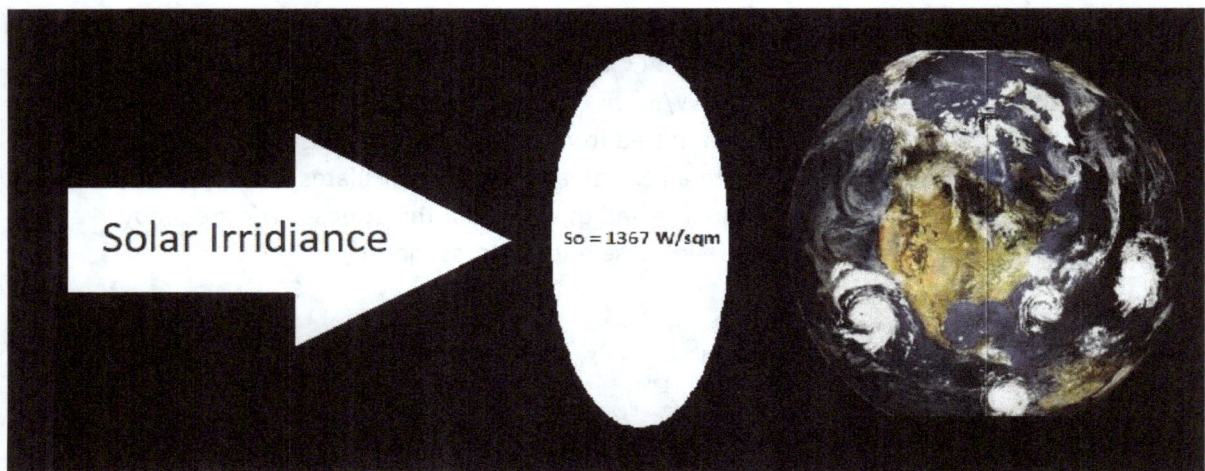

Figure 38: Solar constant; ("Satellite image" was taken from the inside cover of Al Gore's book "An Inconvenient Truth").

Since the IPCC report does not explain in detail how 240W/m² average irradiation power are derived from the solar constant (S_0=1367W/m²), let's see what other "official" climate science institutions reveal about the course of the calculation.

At John F. B. Michell, Meteorological Office, Bracknell, England ("The Green House Effect and Climate Change") I find somewhat more detailed information on the calculation of the temperature of the earth without atmospheric greenhouse effect (Figure 39). He tells us: "The Earth-atmosphere system is warmed by short-wave solar radiation at an average rate of $S_0(1 - α)/4$, where S_0 is the solar constant, α is the fraction of radiation reflected from the Earth and atmosphere, and a **factor of 1/4 is used to account for the spherical shape of the Earth**". Let's work this out:

1367W/m² x (1 - 0.3) / 4 = 239.2W/m²   with α = 0.3 (albedo).

Since the IPCC report states that the result is approx. 240W/m², I assume that this is the "official" method for calculating the greenhouse effect. All other textbooks I know also calculate it this way.

116 • REVIEWS OF GEOPHYSICS / 27,1

2. THE GREENHOUSE EFFECT

2.1. Radiative Effects

The Earth-atmosphere system is heated by solar (short-wave radiation at a mean rate of $S_0 (1 - \alpha)/4$, where S_0 is the solar "constant," α is the fraction of radiation reflected by the Earth and atmosphere, and the factor 4 allows for the spherical geometry of the Earth. This must be balanced by the emission of long-wave (thermal or infrared) radiation to space (Figure 1a). The rate of cooling is given by σT_e^4, where σ is Stefan's constant and T_e is the effective radiating temperature of the system. At equilibrium

$$S_0(1 - \alpha)/4 = \sigma T_e^4 \qquad (1)$$

which assuming the current albedo of 0.30 gives a value of T_e corresponding to 255 K (-18°C). In the absence of an atmosphere, T_e will be the Earth's surface temperature.

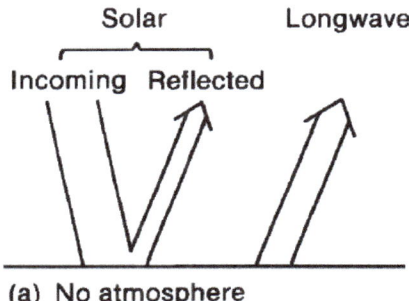

Figure 1. Schematic illustration of the greenhouse effect showing (a) no atmosphere, where long-wave radiation escapes directly to space, and (b) an absorbing atmosphere, where long-wave radiation from the surface is absorbed and reemitted both downward, warming the surface and lower atmosphere, and upward, maintaining radiative balance at the top of the atmosphere.

Figure 39: Source: "The Green House Effect and Climate Change" John F. B. Michell Meteorological Office, Bracknell, England (https://media.gradebuddy.com/documents/460491/43ab9b0f-9274-47b2-86b2-84f25ce20b3d.pdf)

In scientific papers, such trivial calculations are of course not derived in detail and illustrated with a schematic drawing. But in the context of this article written for laypersons, it is possible to deal with self-evident matters in more detail. Therefore, now in detail, derivation and calculation of the atmospheric greenhouse effect in the manner of the IPCC.

As shown in Figure 38, the Earth is hit by solar radiation that passes perpendicularly through a circular area with the radius of the Earth's sphere (r). This circular area is calculated as

$A_k = \pi r^2$.

The solar constant S_0 is used to obtain the power radiated onto the earth.

$P = A_k S_0 = \pi r^2 S_0$

Now we correct with the albedo for the reflected power and get the power absorbed by the earth.

$P = (1 - \alpha) \pi r^2 S_0$

This power is absorbed by the entire surface of the Earth $A_E = 4\pi r2$ (spherical surface) and then emitted again as thermal radiation.

The Stefan-Boltzmann law ($P = A\sigma T^4$) then looks like this (irradiated power = emitted power, in thermal equilibrium).

$P = (1 - \alpha) \pi r^2 S_0 = 4\pi r^2 \sigma T^4$, πr^2 truncates out

$(1 - \alpha) S_0 = 4\sigma T^4$

$\dfrac{(1 - \alpha)S_0}{4\sigma} = T^4$

Resolved after T:

$T = \sqrt[4]{\dfrac{(1-\alpha)S_0}{4\sigma}} = \sqrt[4]{\dfrac{(1-0,3)*1367}{4*5,67*10^{-8}}} \approx 255K \approx -18°C$ Temperature of the earth without greenhouse effect

With the measured average global temperature of about 15°C, this results in an atmospheric greenhouse effect of approx. 33°C. According to this calculation, the earth would be about 33°C colder without greenhouse gases than it really is.

The course of the calculation is now clear. Now let's make a sketch of this model (Figure 40).

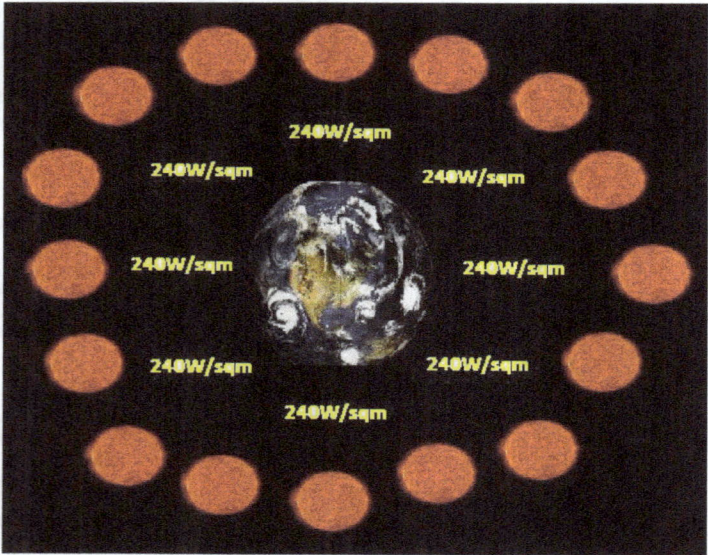

Figure 40:

It is hard to believe, but here the earth is illuminated vertically and uniformly from all directions around the clock with a power of 240W/m². This model makes it possible to bring the degree of difficulty of the calculation to school level. It has nothing whatsoever to do with reality.

Now it is also understandable why IPCC reports do not go into more detail on the determination of the greenhouse effect. It is quite simply really embarrassing when, on the one hand, one pretends to be able to predict the fate of the world's climate with supercomputers, while, on the other hand, one treats the most important basis of the theory of man-made global warming so bunglingly.

IPCC-science does not work with this false model out of ignorance. The problem for the IPCC is that in all reasonably realistic models for calculating the average temperature of the Earth without an atmosphere, the atmospheric greenhouse effect simply disappears.

A bold claim that now needs to be "proven".

A more realistic model for calculating the earth's average surface temperature (without greenhouse effect)

As we have seen from the calculation example with the soccer fields, the Stefan-Boltzmann law does not provide the correct area-averaged average temperature, if the irradiation power is averaged over areas that have different temperatures.

In the IPCC model, the insolation (S_0) is averaged evenly over the entire surface of the Earth. It is therefore to be expected that there is a very large error in this calculation of the mean surface temperature.

To avoid this source of error, a model would have to be developed, that allows the irradiated power to be calculated at every point on the planet's surface. If one knows the radiated power at every point on the planet's surface, one can use the Stefan-Boltzmann law to calculate the surface temperature for every point **without having to average the irradiated power over large areas**. What we are looking for is a surface temperature distribution function $T(\Theta)$ that assigns a temperature to each point on the planet's surface. With this surface temperature distribution function $T(\Theta)$, one can then calculate the mean surface temperature of the planet "practically error-free".

This sounds quite hypothetical. But thanks to the Lunar Diviner Experiment, we have such detailed data on the surface temperature of the Moon, that the surface temperature distribution function $T(\Theta)$ required above, can be determined for the Moon.

With this data, our moon becomes the ideal model for a celestial body without an atmosphere.

When analysing the temperature data from the Lunar Diviner Mission, it turns out, that on the sunny side of the Moon, the surface temperatures are very well described by the Stefan-Boltzmann law when the angle of incidence of the solar irradiation is taken into account (Figure 41).

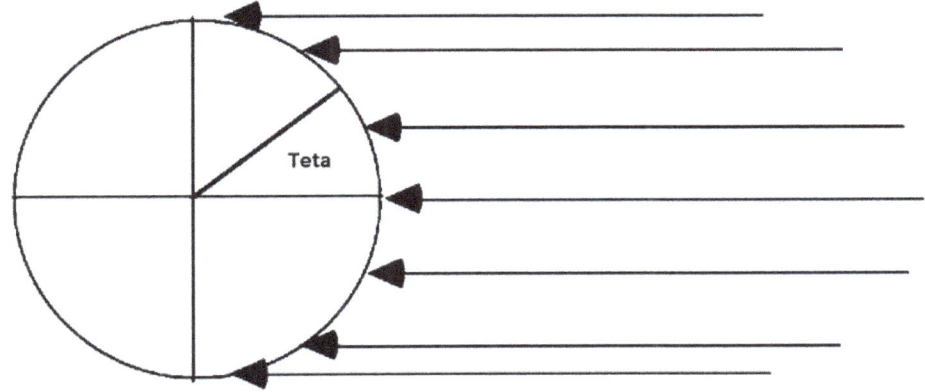

Figure 41: Solar elevation angle Θ (Teta)

According to William et al. 2017
https://www.sciencedirect.com/science/article/pii/S0019103516304869 , the surface temperature distribution function T(Θ) is as follows.

$$T(\theta) = \sqrt[4]{\frac{(1-\alpha)S_0 \cos\theta}{\sigma}}$$ with Θ = solar elevation angle, α = 0.11 (albedo of the moon)

Near the poles, the solar elevation angle Θ (Teta) approaches 90°. This means that the cosΘ approaches zero and very little power is radiated. The surface there is then very cold. At the equator when the sun is at its highest point, Θ = 0°. There, the cosΘ = 1. The full power S_0 is irradiated vertically and the power $(1-\alpha)S_0$ is absorbed. The highest surface temperature is measured here.

Concentric circles with the same angle of incidence and thus the same temperature are thus formed around the point of the sun's highest point (Figure 42).

The shadow side of the moon is not considered further. It simply cools down overnight and is heated up again the next day.

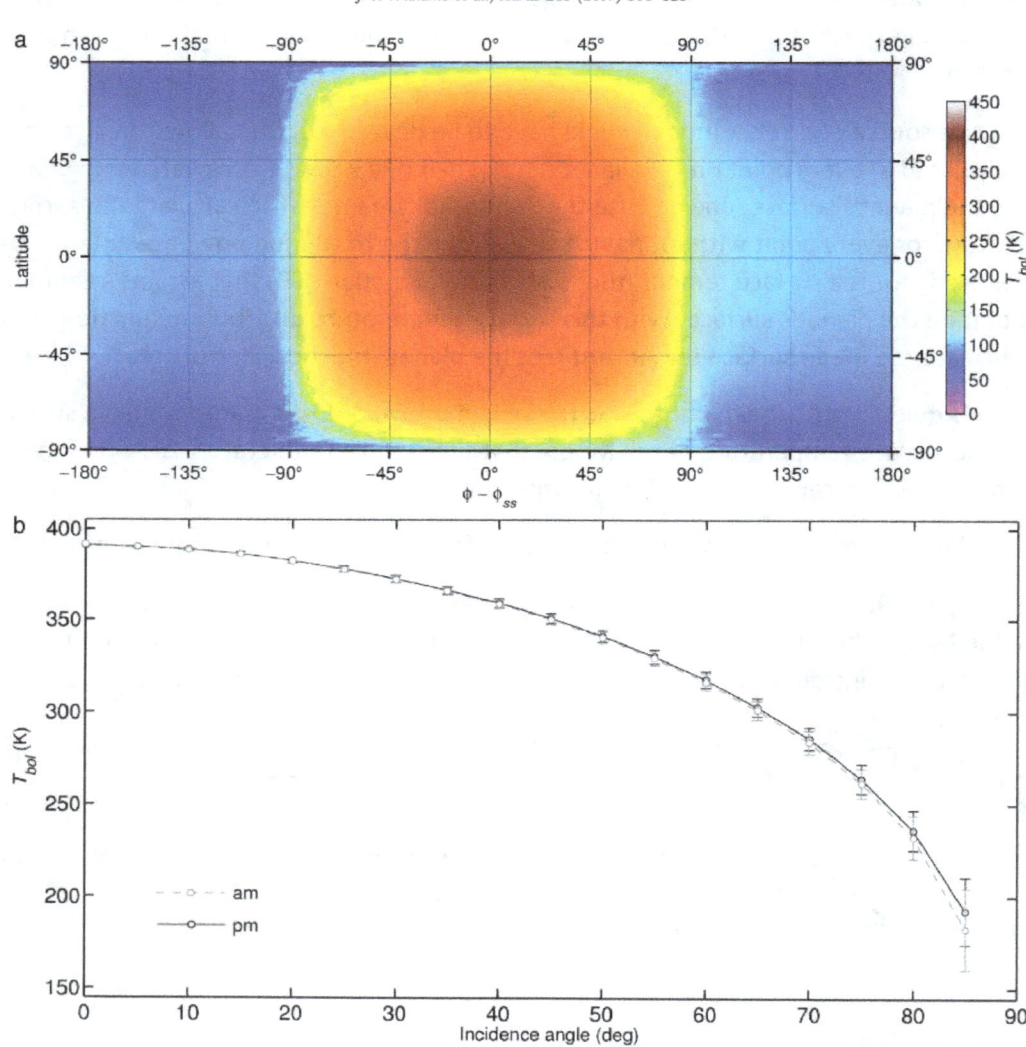

Fig. 10. (a) Average of 24 T_{bol} maps generated with 15° increments of subsolar longitude normalized to the subsolar point (0°, 0°). (b) Mean daytime T_{bol} from (a) for morning hours 6–12 (grey) and afternoon hours 12–18 (black) as a function of incidence angle binned at 5° intervals. Error bars are the standard deviation.

Figure 42: William et al. 2017, Lunar surface temperature (Lunar Diviner Mission),
https://www.sciencedirect.com/science/article/pii/S0019103516304869

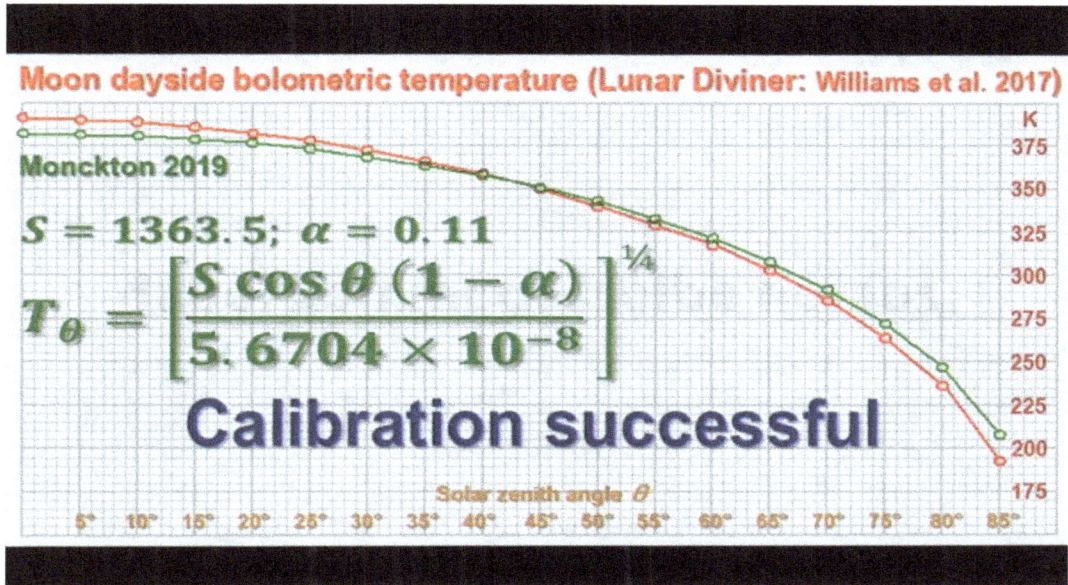

Figure 43: Comparison of measured temperature distribution (red) with calculated distribution (green); source: Lord Christpher Monckton lecture EIKE 2019 Munich

The moon rotates about 28 times slower than the earth. This means that the shady side has much more time to cool down than the night side of the Earth. After sunrise, however, the lunar surface also has more time to heat up. The diurnal variation in temperature is therefore more symmetrical than on Earth. The daily maximum is practically reached at the highest point of the sun (Figure 44).

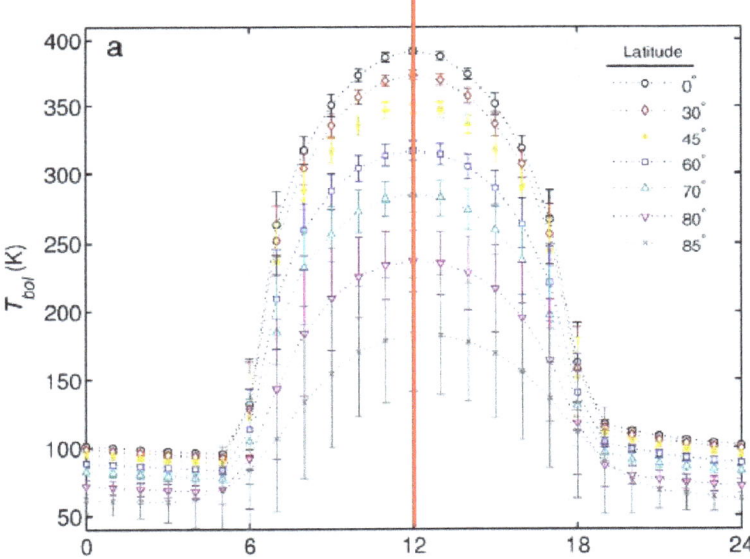

Figure 44: Source: William 2017, Red line marks the solar maximum, https://www.sciencedirect.com/science/article/pii/S0019103516304869

If one compares the diurnal variation of the lunar surface temperature with corresponding terrestrial measurements (Figure 45), one finds that the temperature variations measured on the Earth's surface differ from the lunar values primarily in that the maximum temperature is reached only after the solar maximum (noon) and that the difference between day and night is smaller. These differences are mainly due to the fact, that the Earth rotates about 28 times faster than the Moon and that the heat capacity of the Earth's atmosphere and surface slows down the heating and cooling of the Earth's surface.

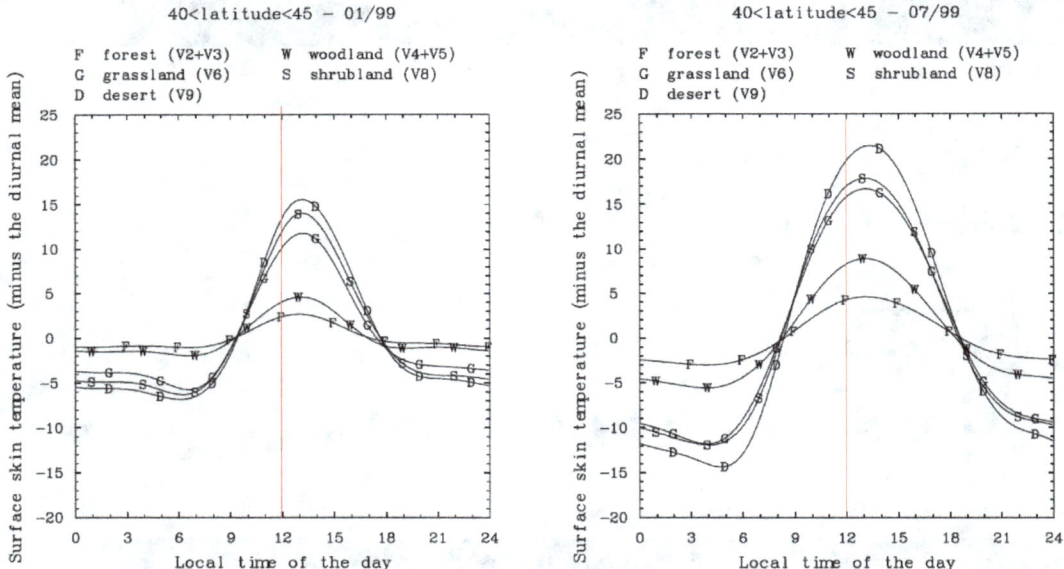

Figure 6. Anomaly of the surface skin temperature diurnal cycle for five surface types for January 1999 (left) and July 1999 (right).

Figure 45: F. Aires et al. 2004, https://pubs.giss.nasa.gov/docs/2004/2004_Aires_ai00100w.pdf

Despite these differences, the diurnal variation of the Earth's surface temperature is much more similar to the diurnal variation of the Moon's surface temperature than to the "IPCC model". Therefore, let's take a look at what comes out when the surface temperature distribution function of the Moon is transferred to the Earth.

When transferring the "moon model" to earthly conditions, we simply replace the albedo of the moon with the albedo of the earth. This means that although we calculate without an atmosphere, the reflection of sunlight on clouds and water surfaces is taken into account.

$$T(\Theta) = \sqrt[4]{\frac{(1-\alpha)S_0 \cos\theta}{\sigma}}$$

with Θ = solar elevation angle, α = 0.3 (albedo of the earth)

This temperature distribution function T(Θ) divides the sunny side of the earth into concentric circles with the same surface temperature (Figure 46).

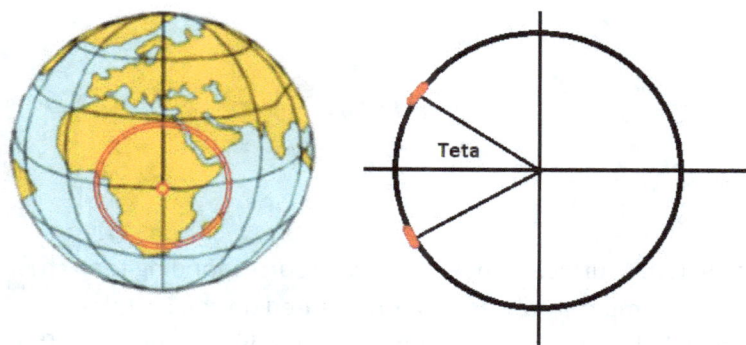

Figure 46: Division of the solar side into concentric rings with the same angle of solar radiation Teta (Θ) according to Uli Weber

Uli Weber describes a numerical solution that calculates the global average temperature via this distribution function. The calculation effort is manageable. Anyone can do the calculation themselves using Excel. The result is 14.03°C and is surprisingly close to the global average temperature of about 15°C "measured" by the IPCC.

A detailed description of the calculation can be found here: "Anmerkungen zur hemisphärischen Mittelwertbildung mit dem Stefan-Boltzmann-Gesetz", (Notes on **hemispheric averaging** with the Stefan-Boltzmann law), Uli Weber https://www.eike-klima-energie.eu/2019/09/11/anmerkungen-zur-hemisphaerischen-mittelwertbildung-mit-dem-stefan-boltzmann-gesetz/

Because it is not particularly difficult in this case, I will also demonstrate a "proper" integral solution for Uli Weber's hemispherical approach.

Here, the concentric ring with the same angle of solar irradiation becomes an infinitesimally narrow area element dA (see Figure 47).

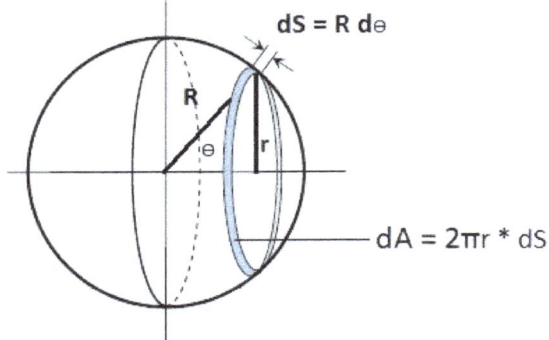

Figure 47:

$$dS = R \, d\theta$$

$$r = R \sin\theta$$

$$dA = 2\pi r \, dS = 2\pi R^2 \sin\theta \, d\theta$$

$$\varepsilon = (1 - \alpha) \approx 0{,}7$$

$$\sigma T^4 = \varepsilon S_0 \cos\theta$$

$$T(\theta) = \sqrt[4]{\frac{\varepsilon S_0}{\sigma}} \sqrt[4]{\cos\theta}$$

The area-averaged mean temperature of the sunlit hemisphere T_m is calculated by forming the area integral of the surface temperature distribution function $T(\theta)$ over the sunlit hemisphere and then dividing by the area of the hemisphere.

$$T_m = \frac{1}{2\pi R^2} \int_{hemisphere} T(\theta) \, dA$$

$$T_m = \frac{1}{2\pi R^2} \int_0^{\frac{\pi}{2}} \sqrt[4]{\frac{\varepsilon S_0}{\sigma}} \sqrt[4]{\cos\theta} \, 2\pi R^2 \sin\theta \, d\theta$$

$$T_m = \sqrt[4]{\frac{\varepsilon S_0}{\sigma}} \int_0^{\frac{\pi}{2}} \sqrt[4]{\cos\theta} \sin\theta \, d\theta = \sqrt[4]{\frac{\varepsilon S_0}{\sigma}} \left| -\frac{4(\cos\theta)^{\frac{5}{4}}}{5} + C \right|_0^{\frac{\pi}{2}}$$

$$T_m = \sqrt[4]{\frac{\varepsilon S_0}{\sigma}} \frac{4}{5} = \sqrt[4]{\frac{0{,}7 * 1368 \frac{W}{m^2}}{5{,}67 * 10^{-8} \frac{W}{m^2 K^4}}} * \frac{4}{5} \approx 288{,}4 K \approx 15°C$$

With an area-averaged average temperature of $T_m \approx 15°C$, the integral solution provides the average temperature of approx. 15°C specified by the IPCC and determined by weather station measurements.

Result: If you let the sun shine on only one side of the Earth, as in real life, and use the results of the Lunar Diviner experiment to calculate the average surface temperature of the Earth, the atmospheric greenhouse effect is simply gone.

I do not want to overstate the result of this calculation, but it does cast doubt on the existence of the atmospheric greenhouse effect.

Accordingly, it does not seem to be the mysterious back-radiation of greenhouse gases that makes the Earth habitable. The Earth's mild climate is more likely due to its rapid rotation (=> short night) and the large heat capacity of its surface and atmosphere. The latent heat that is converted during phase transitions of water also contributes to the high heat capacity of the atmosphere and the Earth's surface.

The fact that the IPCC has to construct an all-round continuous illumination of the earth with 240W/m² in order to be able to show an atmospheric greenhouse effect at all, is for me the best proof of the non-existence of this effect. If the IPCC had a reasonable explanation (or measurement) for the atmospheric greenhouse effect, it would be preached to us daily by every "TV scientist". Instead, the 33°C greenhouse effect is always taken for granted and the derivation of this figure is only mentioned when it cannot be avoided and then only in passing (because it is so embarrassing).

Side note: To take the bitter seriousness out of the matter, in Figure 38 I have used an image of the Earth from climate scientist and Nobel laureate Al Gore's famous book "An Inconvenient Truth". Al has had the satellite image edited a bit to add some drama and emphasis to his message. Every (amateur) meteorologist's eyes are pained by this image. A hurricane has strayed to the equator. And one is even turning the wrong way around off Florida (Figure 48).

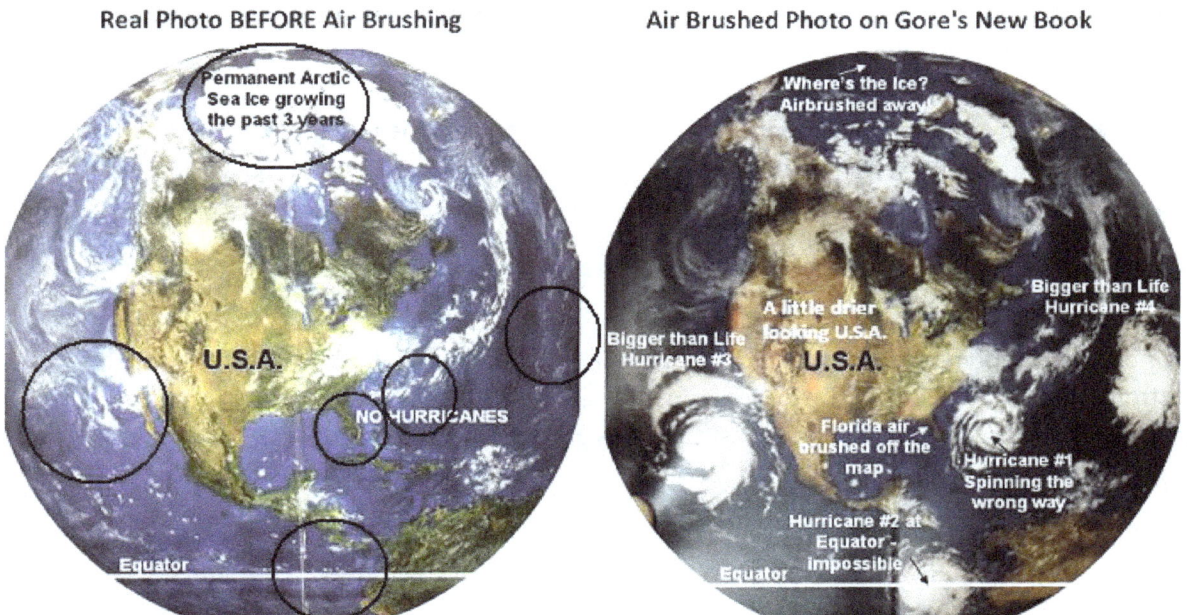

Figure 48: Source: https://australianclimatemadness.com/2010/03/01/more-bad-science-from-the-ipcc/

This picture gives a realistic impression of modern climate science. Don´t take this science too serious. In case you are unable to understand climate science, you should always be aware of the possibility that you can't understand it, because it is stupidly and crudely lied about, or simply wrong.

Greenhouse effect at the molecular level or "Spectroscopy of the Greenhouse effect"

Now that serious doubts have been sown about the existence of the atmospheric greenhouse effect, let us look at how this effect is supposed to come about through the interaction of IR radiation with the gas molecules of the atmosphere. On this level of consideration, too, the reader will come to unexpected insights.

While we have considered energy fluxes of the order of 10^{15} joules per second on a planetary scale, the interaction of IR radiation with individual greenhouse gas molecules converts energies of the order of 10^{-20} joules (per "event").

It has been shown that the description of thermal radiation as a continuous electromagnetic wave is not suitable for "explaining" processes at the molecular level.

Quantum mechanics was developed at the beginning of the 20th century to describe these processes in a useful way.

In quantum mechanics, electromagnetic radiation is not described as a continuous electromagnetic wave, but as a stream of individual light particles, so-called photons. Photons are the carriers of the energy of the radiation. The energy of the photons depends on the wavelength of the radiation. The shorter the wavelength of the radiation, the higher the energy content of its photons.

The absorption of radiation by a molecule is described in quantum mechanics as the interaction of the molecule with a photon of the absorbed radiation.

In order to understand the argumentation of the following chapters, no deep understanding of these processes is necessary. One only has to remember:

- At the molecular level, electromagnetic radiation is described as a stream of individual light particles (photons).
- Photons are the carriers of the energy of radiation.
- The energy of the photons depends on the wavelength of the radiation (the shorter the wavelength of the radiation, the higher the energy content of its photons).
- Molecules can absorb photons with suitable wavelength/energy content.

How to observe/measure the interaction of IR radiation with greenhouse gas molecules?

The most important instrument for studying these interactions is the IR spectrometer. Most IR spectrometers can measure in the wavenumber range from 4000cm^{-1} to 400cm^{-1}. The basic construction and operation of such an instrument is quite simple (Figure 49).

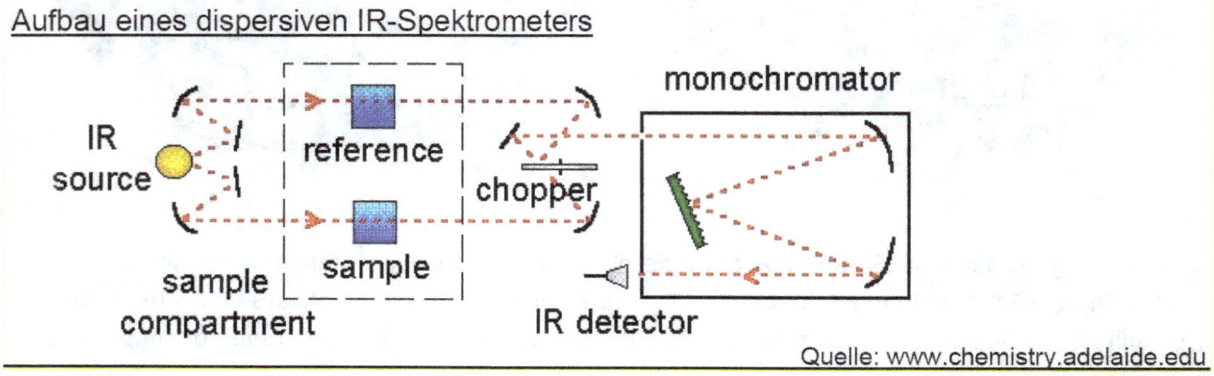

Aufbau eines dispersiven IR-Spektrometers

Figure 49: Source: University of Mainz script: Schematic structure of a double-beam IR spectrometer

In an IR spectrometer, the radiation from an IR source (usually an electrically heated glow rod) is split into two equal beams. One beam is sent through a sample and the other beam is sent through a reference. An empty sample container is often used as a reference. A rotating mirror (chopper) alternately directs the sample beam and the reference beam into the monochromator. The monochromator splits the incident IR radiation into beams of different wavelengths and only allows a certain wavelength to pass through to the detector (similar to how a prism splits visible light into its colours). Because the rotating mirror alternately directs IR radiation from the sample and the reference into the monochromator, it is possible to compare the transmittance of the sample with the transmittance of the reference in the detector. To record a spectrum, the monochromator is adjusted so that it scans all wavelengths from 400cm^{-1} to 4000cm^{-1} in succession. In parallel, the detector records the transmittance of the sample compared to the reference at the respective wavelength.

IR spectrometers provide a diagram showing the transmittance of a sample for IR radiation as a function of the wavelength of the IR radiation. Such a representation is called an IR spectrum. Figure 49 shows an IR spectrum of CO_2 gas.

The transmittance of a sample (formula symbol: τ) indicates the fraction of radiation that can penetrate the sample.

Transmittance τ = 1 means that the sample is completely transparent to light. τ = 0.5 means that half of the incident light passes through the sample. The rest is absorbed. τ = 0 means that the sample is impermeable.

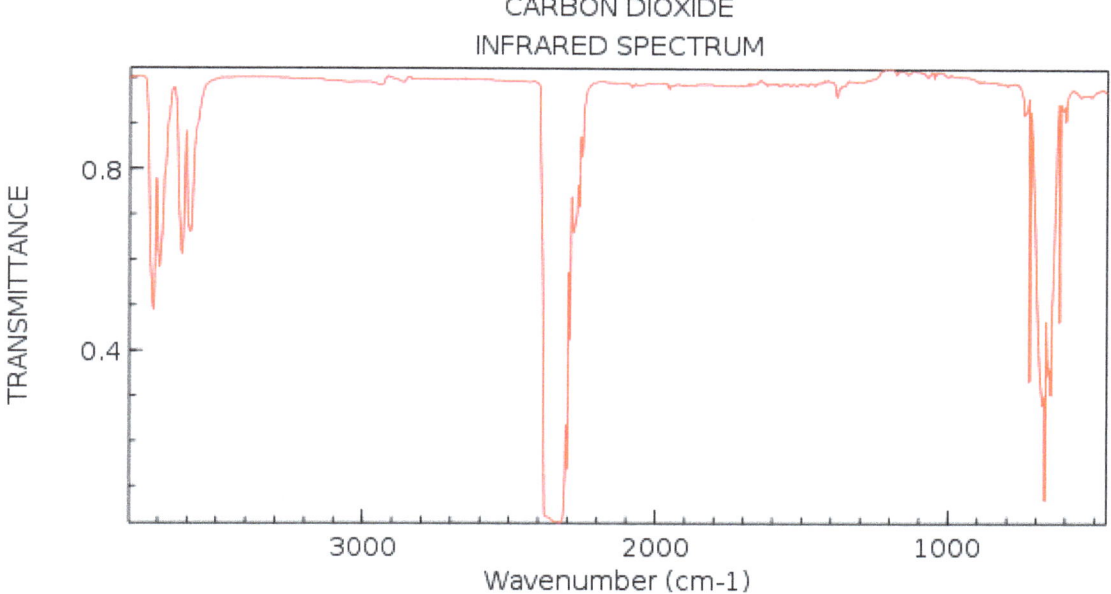

Figure 50: IR spectrum of CO_2. Source: NIST

In IR spectroscopy, it is common to specify the wavenumber instead of the wavelength. The wavenumber indicates how many wavelengths fit onto one centimeter. The wavenumber has the unit cm^{-1}.

In the IR spectrum of CO_2 shown above (Figure 50), one can clearly see that in certain areas the transmittance is very low. For the climate discussion, we are only interested in the strong absorption (or low transmittance) at the wavenumber approx. $666 cm^{-1}$. Converting this wavenumber into wavelength results in a wavelength of λ = 0.01m/666 = 0.000015m = 15µm.

With the help of the Plank equation and the wavelength of the absorbed photons, we can get an idea about how much energy is transferred during such an absorption process. For the case of the 15µm absorption of CO_2 we get.

$E = \frac{hc}{\lambda} = h\nu$ with: h = 6.626069 · 10^{-34} J sec Planck's constant, c: Speed of light (in vacuum) c = 299792458m/s ≈ 3 · 10^8 m/s , λ: wavelength, ν: frequency

$$E = \frac{6{,}626069 * 10^{-34} J\,s * 3 * 10^8 m/s}{15 * 10^{-6} m} \approx 1.3 * 10^{-20} J$$

What happens when CO_2 interacts with infrared light?

Literature on this: H. Hug, *Chemische Rundschau,* 20 Feb, p. 9 (1998) ; 10 Aug 2012 The anthropogenic greenhouse effect - a spectroscopic trifle by Heinz Hug https://www.eike-klima-energie.eu/wp-content/uploads/2016/12/Hug-pdf-12-Sept-2012.pdf .

As already noted, greenhouse gas molecules can absorb photons of IR radiation. For the case of CO_2, let's take a closer look at this process.

CO_2 is a rod-shaped molecule in which a carbon atom is bonded to two oxygen atoms (Figure 51). To understand the vibrational behaviour of CO_2, the bonds between the atoms can be thought of as similar to spiral springs. The elastic properties of the bonds allow the molecule to perform bending vibrations and stretching vibrations (valence vibrations). In the CO_2 molecule, the carbon atom (in the middle of the molecule) carries a weak positive charge. The oxygen atoms at the ends of the molecule carry weak negative charges.

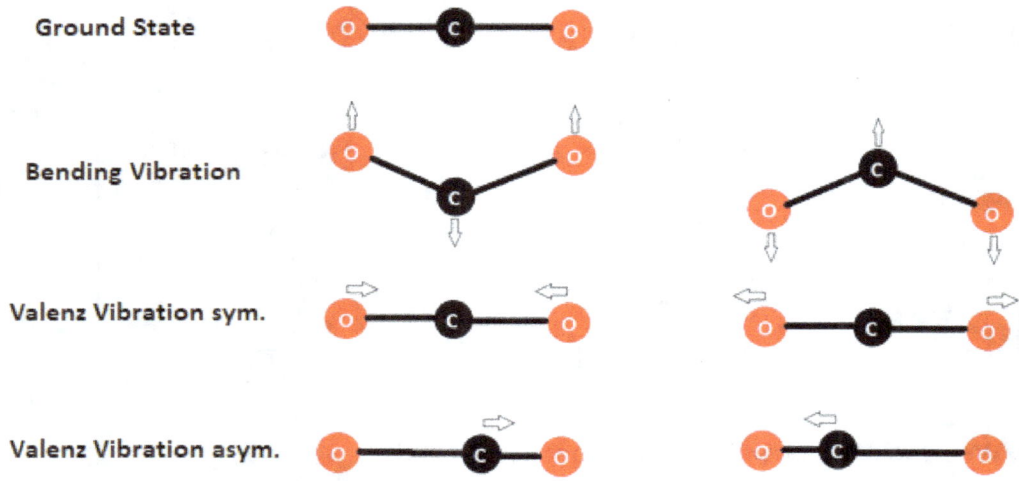

Figure 51: Oscillation possibilities of the CO_2 molecule. Bending vibration: absorbs at 666cm^{-1}, the abbreviation (01^{1}0) is often used as notation for this state of vibration; the symmetric valence vibration: is not IR active (Raman at 1366cm^{-1}), the asymmetric valence vibration: absorbs strongly at 2349cm^{-1}.

In the case of the bending oscillation and the asymmetric valence oscillation, the centre of charge of the molecule shifts. I.e. these oscillations generate changing electric fields similar to a rod antenna. Therefore, CO_2 molecules should also be able to absorb and re-emit electromagnetic radiation (thermal radiation).
So it does. But while a mass vibrating on a spiral spring can absorb and release a wide range of energies, each type of vibration of the CO_2 molecule can only absorb or release very specific amounts of energy. This is called energy quantisation. More about this can be found in spectroscopy textbooks such as Hesse/Meier/Zeeh.
In the IR spectrum of CO_2, the wavelengths/wavenumbers at which the CO_2 molecules absorb energy from the irradiated IR radiation can be recognised by a decreased transmittance of the sample. The asymmetric stretching oscillation of CO_2 absorbs at a wave number of 2349cm^{-1} in a range in which the atmosphere is impermeable to IR radiation anyway. It therefore plays no role in the climate discussion.
Only the bending vibration of CO_2 is relevant for the climate discussion. It absorbs at wave number 666cm^{-1} (i.e. at a wavelength of approx. λ = 15µm) and lies in a range in which the atmosphere, according to the IPCC, is not completely impermeable.
Figure 51 shows the 666cm-1 absorption band in high resolution. To the left and right of this band are several small absorptions that result from the fact that the bending vibration of the CO_2 molecule is superimposed by rotational movements of the CO_2 molecule (so-called rotational bands).

To show how the presence of other gas molecules affects the absorption behaviour of CO_2 molecules, three CO_2 spectra with different gas admixtures were superimposed in Figure 52. In all three spectra, the same number of CO_2 molecules was in the beam path of the spectrometer. Pure CO_2, CO_2 diluted with helium and CO_2 diluted with nitrogen were measured. It **can be seen that the strength of the CO_2 absorption band depends strongly on which gases are contained in the sample in addition to CO_2.**

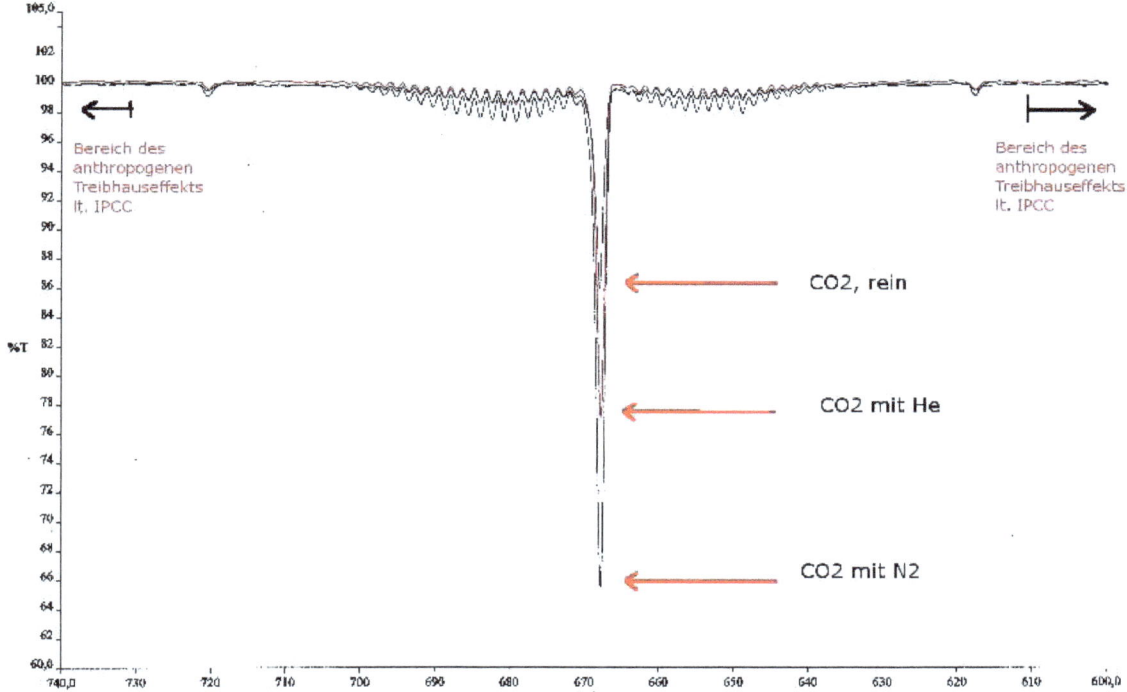

Figure 52: Source H.Hug: CO_2 absorption band at 666cm^{-1} measured in pure CO_2, in mixture with helium and in mixture with nitrogen. In all measurements, the same amount of CO_2 molecules is in the sample, https://www.eike-klima-energie.eu/wp-content/uploads/2016/12/Hug-pdf-12-Sept-2012.pdf

This effect is of central importance for heat transport in the Earth's atmosphere. To understand how this effect comes about, we need to discuss the following processes:
- Absorption and emission of 15µm IR radiation by CO_2
- Absorption and thermalisation of 15µm IR radiation by CO_2
- Thermal excitation and emission of 15µm IR radiation by CO_2

Absorption And Emission of 15µm IR Radiation by CO2

When a CO_2 molecule is hit by a photon with a wavelength of 15µm, it can absorb the energy of this photon. In doing so, it passes from its ground state, in which it does not oscillate, to a state of higher energy (excited state, often abbreviated as (01^10)), in which it performs the bending oscillation described above. In this excited state, the CO_2 molecule stores the energy of the absorbed photon in the form of vibrational energy (see energy diagram Figure 53). This excited state is stable for a short period of time. When the CO_2 molecule returns to its ground state, it emits a photon again and stops oscillating. The emission of the photon occurs randomly in any spatial direction.

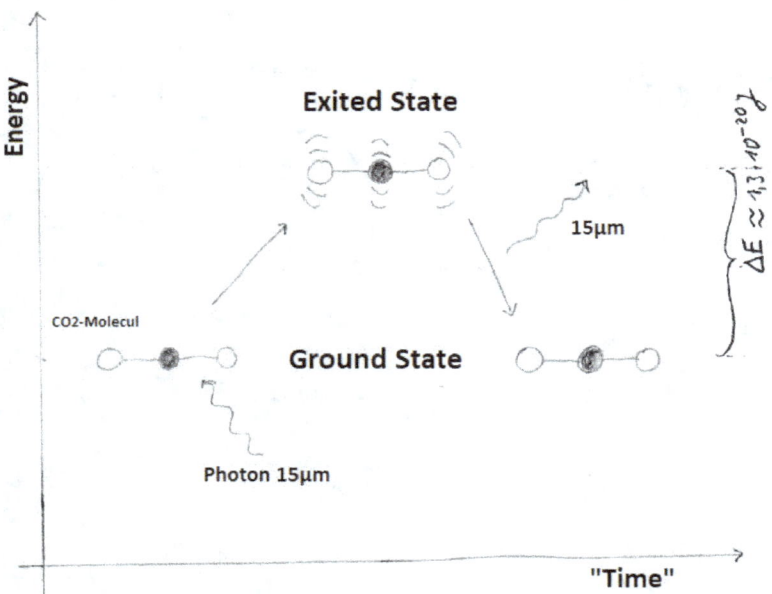

Figure 53: Energy diagram, excitation of a CO_2 molecule

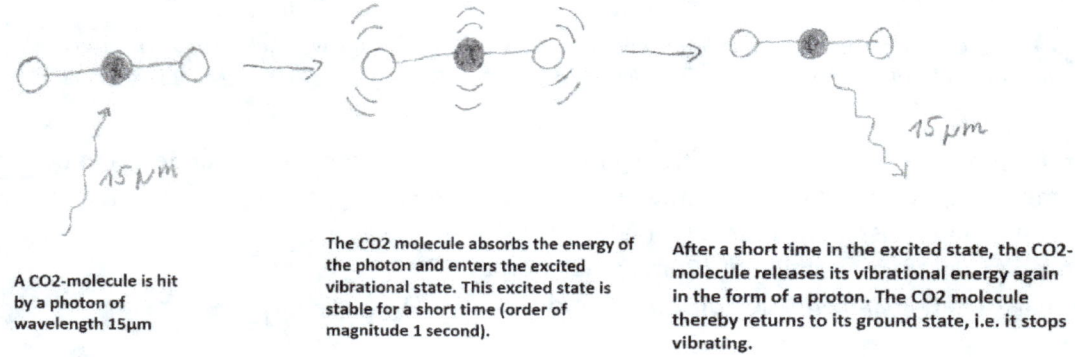

Figure 54: Absorption and emission of a photon by CO_2

It is important to note that CO_2 can only absorb 15μm radiation from its ground state. The excited state cannot absorb 15μm radiation.

A quantum mechanical description of such processes can be found in A. Einstein, Physikalische Zeitschrift 18, 121, 1917 The Quantum Theory of Radiation; http://web.ihep.su/dbserv/compas/src/einstein17/eng.pdf . The HITRAN database contains the constants that describe this process https://www.spectralcalc.com/spectral_browser/db_data.php . For the strongest absorption of CO_2, a decay constant of $K_d = 1.542 s^{-1}$ is given there, which results in a half-life-time of the excited CO_2-(01^10) oscillatory state of approx. 0.45s. Thus a "lifetime" of about 0.1sec to 1sec is realistic for back of the envelop calculations (the calculation of this half-life is demonstrated below).

Absorption and Thermalisation of 15μm IR Radiation by CO_2

When a CO_2 molecule in the excited state collides with another gas molecule, its vibrational energy can be transferred to the colliding gas molecule (Figure 55). After this collision, the CO_2 molecule has returned to its ground state. It can then absorb a suitable photon again.

The molecule that collided with the excited CO_2 thereby converts the energy transferred to it into kinetic energy. This means that the gas molecule involved in the collision with the excited CO_2 molecule picks up speed in the process.

Since temperature is a measure of how much kinetic energy the molecules of a gas have, the gas becomes warmer during this process.
Thermal radiation is thus converted into molecular movement, i.e. heat. This conversion of radiation energy into heat is also called "**thermalisation**".

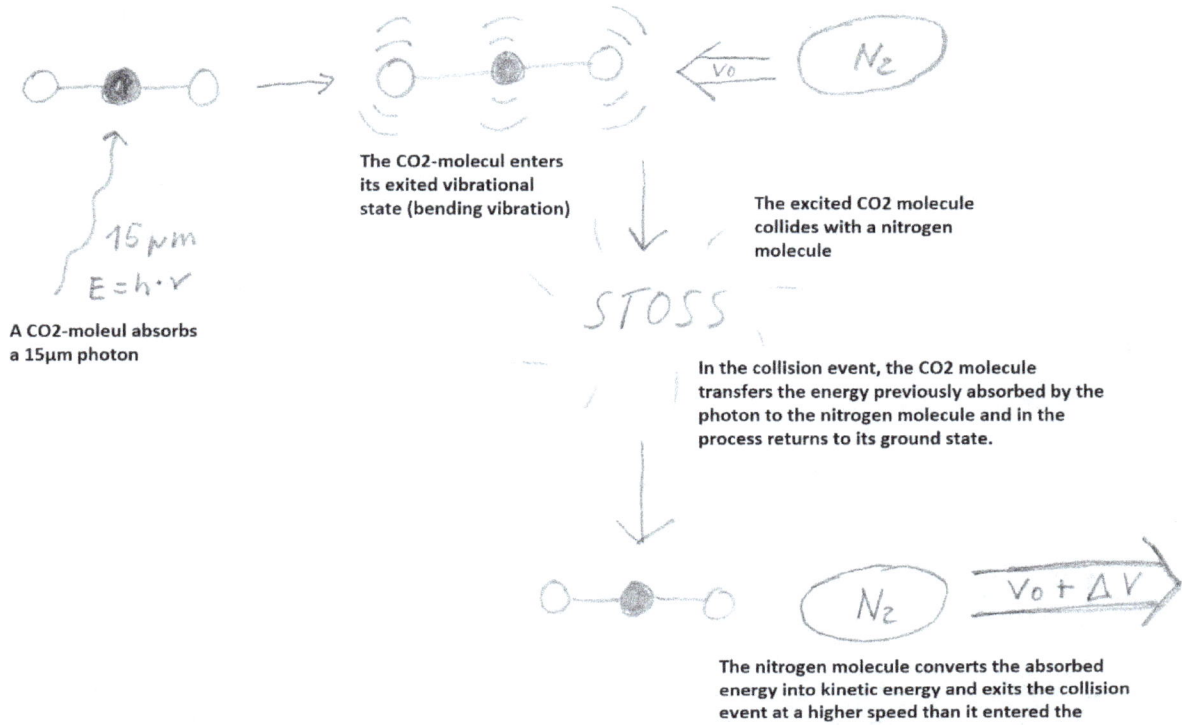

Figure 55: Thermalisation

Thermal Excitation and Emission of 15µm IR Radiation by CO_2

The reverse process is also possible. A CO_2 molecule in the ground state absorbs so much energy in a collision with another gas molecule that it goes into the excited state. It can then emit a photon and return to its ground state. During this process, the kinetic energy of the gas molecules is converted into thermal radiation. **The gas cools down in the process**. This process is called **thermally excited emission** (Figure 56).

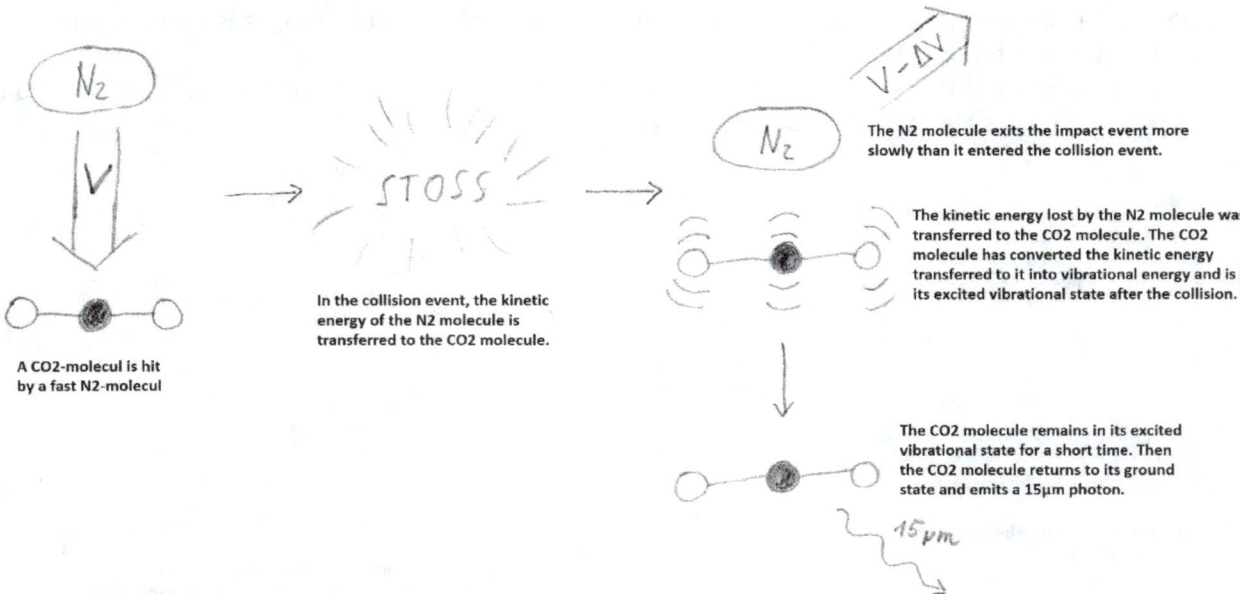

Figure 56: Thermally excited emission

Other greenhouse gas molecules, such as water, methane, sulphur hexafluoride, etc., behave in the same way.

Because it is so important, I will briefly summarise again how greenhouse gas molecules can interact with IR radiation of the appropriate wavelength.

1. **Greenhouse gas molecules can absorb IR radiation with a suitable wavelength, store the energy absorbed in the process for a short time and then emit it again as IR radiation in any direction.**
2. **Greenhouse gas molecules can absorb IR radiation with a suitable wavelength and then transfer the absorbed energy to other gas molecules. This process heats the air (thermalisation).**
3. **Greenhouse gas molecules can absorb energy when they collide with other air molecules and then emit the energy transferred to them as IR radiation. In the process, the air can cool down (thermally excited emission).**

Now it is understandable why CO_2 absorbs more strongly when mixed with other gases than in its pure form. In gas mixtures, CO_2 molecules, after they have absorbed IR radiation and are in an excited state, transfer their vibrational energy to other gas molecules. This energy transfer to other gas molecules opens up an alternative, very fast way for the CO_2 molecules to return to their ground state, in addition to the emission of IR radiation. Therefore, the proportion of CO_2 molecules in the ground state is greater in gas mixtures than in pure CO_2. Accordingly, a stronger CO_2 absorption is observed in gas mixtures than in pure CO_2.

In pure CO_2, a CO_2 molecule in the excited state can only transfer its energy to another CO_2 molecule, which in turn then can enter the excited state. In total, the proportion of CO_2 molecules in the ground state does not increase via such an energy transfer. Therefore, the absorption band in pure CO_2 is weaker than in mixtures with other gases (Figure 51).

"IPCC greenhouse effect" at the molecular level

Now that we have a rough idea of how thermal radiation and gas molecules interact, let's look at how the atmospheric greenhouse effect is **supposed to** work, according to the IPCC (Figure 57).

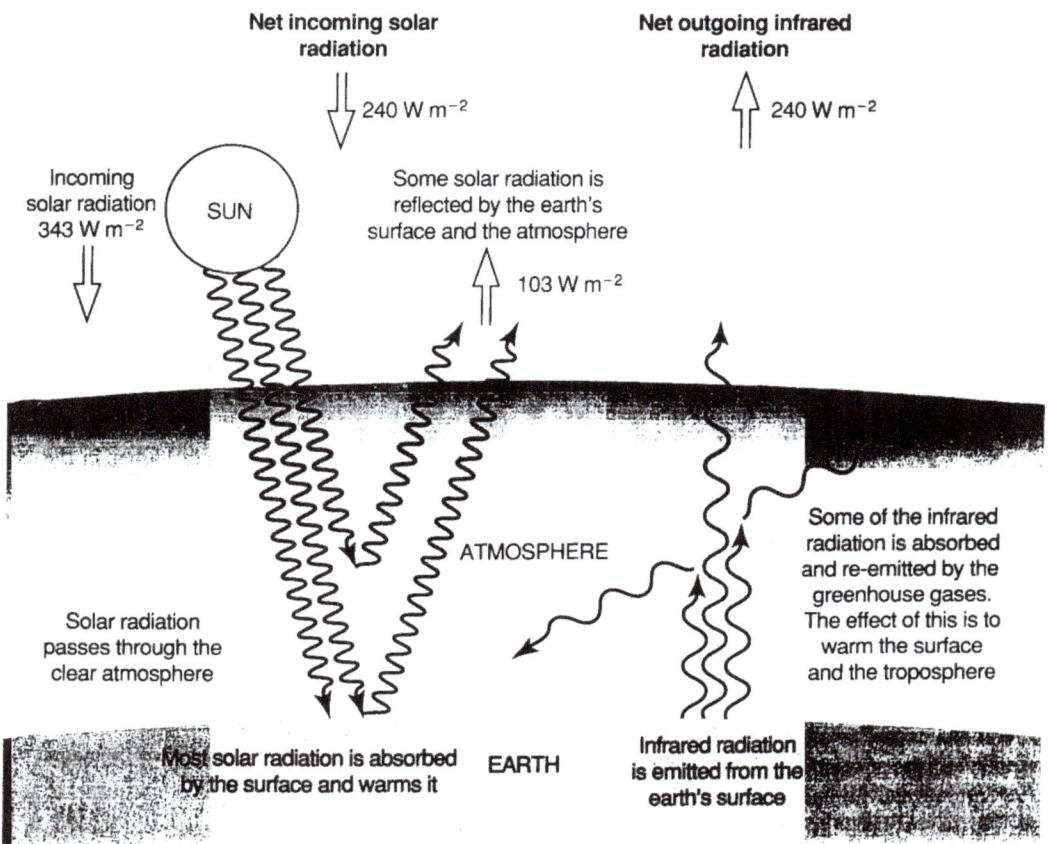

Figure 1: A simplified diagram illustrating the global long-term radiative balance of the atmosphere. Net input of solar radiation (240 Wm⁻²) must be balanced by net output of infrared radiation. About a third (103 Wm⁻²) of incoming solar radiation is reflected and the remainder is mostly absorbed by the surface. Outgoing infrared radiation is absorbed by greenhouse gases and by clouds keeping the surface about 33 °C warmer than it would otherwise be.

Figure 57: Source: IPCC Report "Climate Change 1994 Radiative Forcing of Climate Change and An Evaluation of the IPCC IS92 Emission Scenarios", page 15 https://www.ipcc.ch/site/assets/uploads/2018/03/climate_change_1994-2.pdf

Short-wave sunlight penetrates the atmosphere and warms the Earth's surface. The warm surface of the Earth emits part of the energy radiated onto it as thermal radiation (IR radiation) in the direction of outer space.

Some of this thermal radiation escapes through the atmosphere into space, cooling the Earth. Another part of the thermal radiation is absorbed by greenhouse gas molecules and then re-radiated in a random direction. This re-radiation in a random direction causes about half of the previously absorbed IR radiation to be re-radiated back to the Earth's surface and into the lower layer of the atmosphere (troposphere). **This <u>re-radiation</u> heats the Earth's surface and troposphere.** *The other half can escape into space.*
If we now increase the concentration of greenhouse gas molecules in the atmosphere, the proportion of IR radiation increases, which is absorbed by greenhouse gases on its way into space and radiated back to Earth. This increases the greenhouse effect and makes the Earth warmer.

It is very important to note that global warming is supposed to be caused by the <u>back radiation</u> of greenhouse gases on the earth's surface and troposphere and not somehow by "warm air".

Actually, this sounds quite plausible. But after what was said before about the radiationless deactivation of excited states (thermalisation), one can guess that this model does not quite correspond to reality.
Measurements of the atmosphere's permeability to IR radiation also gives a different picture.

Saturation in the range of the 15μm radiation

Anyone who has tried to paint a black wall white has made the following observation: After the first paint, the wall is dark grey. After the second coat it is grey. With the following coats of paint, the wall becomes lighter and lighter. Then a coat thickness is reached where a further coat of paint does not make the wall any whiter. The first coats of paint make the wall much whiter. With the last coats, you have to look closely to see a difference from the previous coat.

A similar effect is also observed when the concentration of a greenhouse gas in the atmosphere is increased. I will explain this effect using CO_2 as an example (Figure 58). If the atmosphere contains no CO_2, the IR frequencies that CO_2 absorbs can escape unhindered into space (Figure 58/1). If you now add a small amount of CO_2 to the atmosphere, each CO_2 molecule is hit by the IR radiation and can absorb radiation (Figure 58/2). If you increase the CO_2 concentration further, it happens that some CO_2 molecules move in the "shadow" of other CO_2 molecules. At this CO_2 concentration, each additional CO_2 molecule no longer increases the absorption effect (Figure 58/3). If the CO_2 concentration is increased further, a CO_2 concentration is reached at which practically complete absorption is achieved. A further increase in the CO_2 concentration then causes practically no further increase in absorption (Figure 58/4). The atmosphere is then practically impermeable to the IR radiation absorbed by the CO_2. This state is called saturation. When the saturation concentration of CO_2 is reached, a further increase in CO_2 concentration does not lead to any measurable increase in IR absorption (and thus the greenhouse effect).

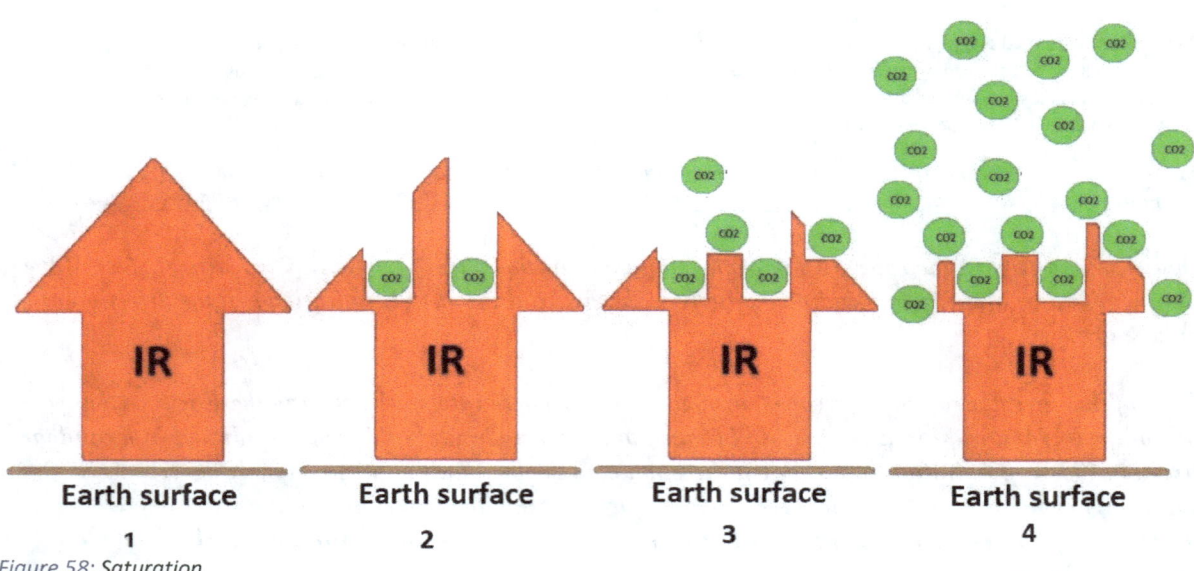

Figure 58: Saturation

This dependence of the permeability of a medium from the concentration of an absorbing component of the medium is described by Lambert-Beer's law (from 1852, so nothing new). It reads:

$E_\lambda = \log 10(I_0/I) = \varepsilon_\lambda \, C \, d$ with E_λ: extinction at the wavelength λ, I_0: irradiated light intensity, I: intensity that passes through the medium, ε_λ: extinction coefficient for the wavelength λ, C: concentration of the absorbing substance, d: layer thickness of the medium that the light has to pass through

Or as an exponential function:

$$\tau = \frac{I}{I_0} = 10^{-\varepsilon_\lambda C\, d} \qquad \text{with } \tau\text{: transmission, see graph Figure 58}$$

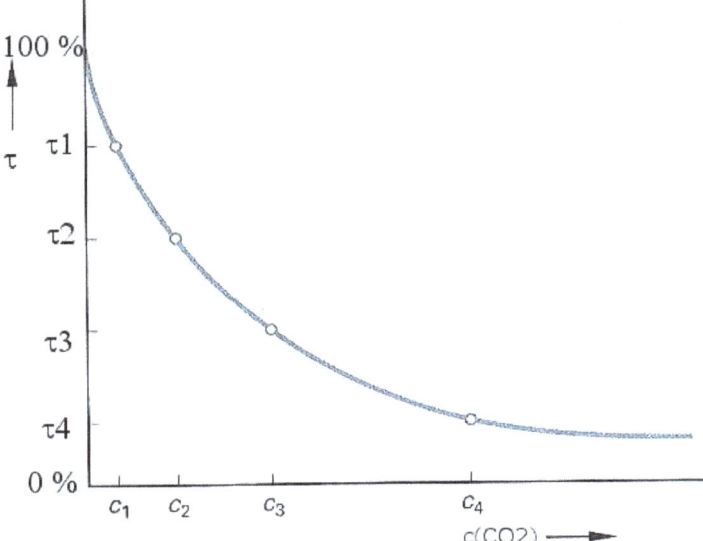

Figure 59: Source H.Hug: Dependence of the transmission ("percentage permeability") of air on the CO_2 concentration, https://www.eike-klima-energie.eu/wp-content/uploads/2016/12/Hug-pdf-12-Sept-2012.pdf

As can be seen in graph Figure 59, the transmittance hardly changes once certain CO_2 concentrations are exceeded. This means that the atmosphere very quickly becomes impermeable to IR radiation with a wavelength of 15μm. A further increase in the CO_2 concentration then only shortens the range of the 15μm radiation in the atmosphere.

Heinz Hug has determined the extinction coefficient ε_λ by laboratory measurements. In air samples with 357ppm (0.0159mol/m³) of CO_2 and 2.6% of water he obtains:

$\varepsilon_{15\mu m}$ = 20.2 m²/mol

With this extinction coefficient, the transmittance for a distance of 10m in air is

$$\tau = \frac{I}{I_0} = 10^{-20{,}2\frac{m^2}{mol} \,*\, 0{,}0159\frac{mol}{m^3} \,*\, 10m} = 10^{-3{,}21} = 0{,}0006$$

This means that after passing through approx. 10m of air, IR radiation with a wavelength of 15μm is 99.94% absorbed. It can therefore be assumed that the atmosphere in the area of the 15μm band of CO_2 is practically impermeable. A further increase in the CO_2 concentration can therefore not cause a further increase in absorption. That means, that a CO_2-greenhouseeffect (so it exists at all) is already saturated at the current atmospheric CO_2 concentration.

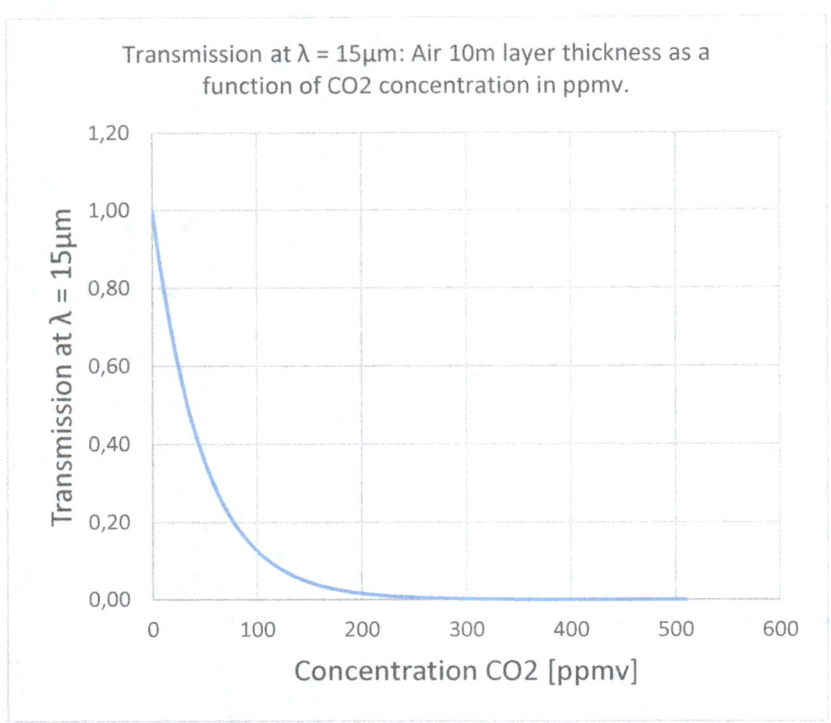

Figure 60: Transmission (permeability) of a "near-ground" 10m thick air layer as a function of CO_2 concentration, calculated with the $\varepsilon_{15\mu m} = 20.2\ m^2/mol$ determined by Hug.

As can be seen in Figure 60, even at the much-cited pre-industrial 280 ppmv of CO_2, the 15μm absorption is at saturation.

Back radiation of 15μm radiation from the atmosphere to the Earth's surface is therefore limited to a ground layer measuring only a few metres (see Figure 61).

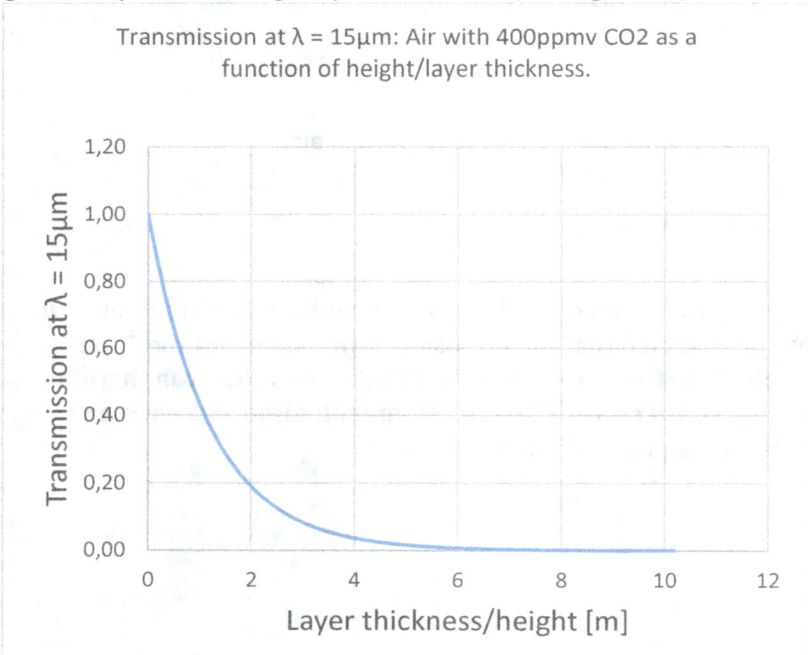

Figure 61: Transmittance of a "near-ground" air layer as a function of layer thickness or height, calculated with the $\varepsilon_{15\mu m} = 20.2\ m^2/mol$ determined by Hug.

IPCC Trying to save the greenhouse effect

The IPCC is aware of this problem and therefore argues, that the rotational bands above and below the 15µm band (see Figure 51) provide an amplification of the greenhouse effect with a further increase in the atmospheric CO_2 concentration. Since the rotational bands only absorb very weakly, even the IPCC cannot construct a large greenhouse effect from these absorptions. According to the IPCC, the CO_2 currently contained in the atmosphere should cause a back radiation from the atmosphere to the Earth's surface of about $32W/m^2$. In the case of a doubling of the current CO_2 concentration, the IPCC estimates that the back radiation caused by CO_2 will increase by about $4W/m^2$ (see Figure 62). The nice round figure $4W/m^2$ suggests that this figure is an estimate rather than an experimentally determined quantity. As we shall see, it is not at all important how accurate this figure is. When calculating the global warming caused by this, the fourth root is taken from it anyway and its warming effect almost disappears.

16

For example, an increase in atmospheric CO_2 concentration leads to a reduction in outgoing infrared radiation and a positive radiative forcing. **For a doubling of the pre-industrial CO_2 concentration, in the absence of any other change, the global mean radiative forcing would be about 4 Wm^{-2}. For balance to be restored, the temperature of the troposphere and of the surface must increase, producing an increase in outgoing radiation. For a doubling of CO_2 concentration, the increase in surface temperature at equilibrium would be just over 1 °C, if other factors (e.g., clouds, tropospheric water vapour and aerosols) are held constant.** Taking internal feedbacks into account, the 1990 IPCC report estimated that the increase in global average surface temperature at equilibrium resulting from a doubling of CO_2 would be likely to be between 1.5 and 4.5 °C, with a best estimate of 2.5 °C.

Figure 62: Source: IPCC Report, Climate Change 1994 Radiative Forcing of Climate Change And An Evaluation of the IPCCIS92 Emission Scenarios, page 16 https://www.ipcc.ch/site/assets/uploads/2018/03/climate_change_1994-2.pdf

Because it does have a certain entertainment value, let's calculate, with the official IPCC figures, what warming effect a doubling of the atmospheric CO_2 concentration should have according to the IPCC. To do this, we enter the Stefan-Boltzmann equation with the average temperature of the Earth "determined" by the IPCC (15°C or 288K).

$P/A = \sigma T^4 = 5.67 \times 10^{-8}$ $Wm^{-2}K^{-4}$ $\times 288^4$ $K^4 = 390 W/m^2$ with $\sigma = 5.670 \times 10^{-8}$ $Wm^{-2}K^{-4}$

With this average temperature, the average irradiation of the earth is $390W/m^2$.
We add the additional back radiation of $4W/m^2$ caused by the doubling of the CO_2 concentration to the $390W/m^2$ because the additional radiated power must also be emitted in thermal equilibrium.

$$T = \sqrt[4]{\frac{P}{\sigma A}} = \sqrt[4]{\frac{394}{5,67*10^{-8}}} K = 288,7K = 15.7°C$$

Using the official IPPC figures/assumptions, the doubling of the atmospheric CO_2 concentration causes an increase in the world average temperature of 0.7°C.

0.7°C increase in the global average temperature really doesn't sound like a climate crisis or catastrophe. And even without a UN world government, humanity will cope well with this warming.

Water vapour feedback

For the UN, apart from infectious diseases and perhaps an alien attack at some point, there are few alternatives to CO_2. Whoever controls CO_2 emissions controls the world. To save the CO_2 story, a positive feedback through water vapour has been invented. This feedback is supposed to amplify the rather meagre direct warming effect of the CO_2 doubling to 2.5 to 4.5°C (see Figure 61).

This is supposed to work as follows:
1. Increase in CO_2 concentration causes a small increase in the "world average temperature".
2. The small increase in the average world temperature causes more evaporation of water (of which there is a lot in the world) and thus a higher concentration of water vapour in the atmosphere.
3. Water vapour is a greenhouse gas. More water vapour in the atmosphere causes even more back radiation and thus "greenhouse warming".
4. The now higher air temperature allows the atmosphere to absorb even more water (Clausius-Clapeyron equation).
5. And so forth....

The IPCC report, CLIMATECHANGE 2001: THE SCIENTIFIC BASIS, *7.2.1 Physics of the Water Vapour and Cloud Feedbacks*, describes this feedback in some detail.
https://www.ipcc.ch/site/assets/uploads/2018/03/TAR-07.pdf

The attentive reader already suspects how this must end. **According to the IPCC, the earth's climate is a very unstable system that reacts to small disturbances with a "runaway global warming".**

Does this positive feedback through water vapour really exist?

The ice core data do not convey the picture of a positive feedback (Figure 23). If this positive water vapour feedback really existed, one would have to be able to observe a "runaway global warming" in the ice core data at the end of each ice age. This is not the case.
Moreover, it is impossible to understand how a climate system, that contains strong positive feedbacks, can be relatively stable over thousands of years.
Experience even teaches the opposite. On days with high humidity there is usually more cloud formation. Cloud formation very strongly dampens the temperature rise and usually causes cooling. **In other words, the IPCC wants to sell us a negative feedback that dampens the temperature rise as a positive feedback that leads to climate catastrophe.**

With reference to the saturation effect, one should actually end the CO_2 discussion immediately. But because this is not about CO_2 but about power, IPCC climate science constructs a radiative transfer mechanism that somehow allows the heat radiation absorbed by CO_2 to escape from the Earth's surface into space or to be radiated back from the atmosphere towards the Earth's surface. This mechanism is described by the radiative transfer equation.

Radiative transfer equation

How this is supposed to work can be seen in corresponding textbooks (e.g. Radiative Transfer In The Atmosphere And Ocean, Knut Stamnes, Gary E. Thomas, Jakob J. Stamnes). A clear derivation and integration of this equation by Ross Bannister can be found under this link:
http://www.met.reading.ac.uk/~ross/Science/RadTrans.pdf.

I will only superficially describe here how one tries to "save" the greenhouse effect with the help of the radiative transfer equation.

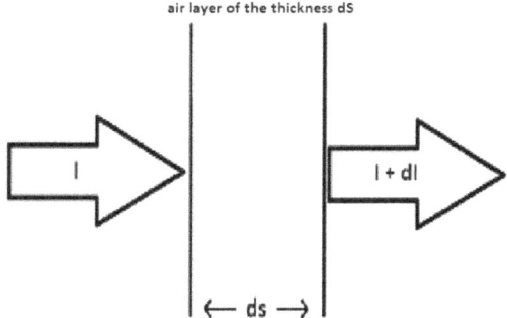

Figure 63: Radiative transfer through the atmosphere

We consider a light beam with the wavelength λ and the specific intensity I_λ emitted from the Earth's surface towards space (Figure 63).
On its way through the atmosphere, the beam is amplified by emission of the greenhouse gases contained in the air. At the same time, it is weakened by absorption processes and scattering.
For the change of the specific intensity dI_λ on the distance ds we get.

$\frac{dI_\lambda}{ds}$ = $specific\ Emission\ of\ the\ Wavelength\ \lambda - specific\ absorption\ respectively\ scattering\ of the\ Wavelength\ \lambda$

$$\frac{dI_\lambda}{ds} = j_\lambda - \varepsilon_\lambda I_\lambda$$

This simple relationship is called the radiative transfer equation.

The solution of the radiative transfer equation is not entirely trivial and opens up the possibility of incorporating premises that allow radiative transfer at the "IR greenhouse gas frequencies" through the atmosphere into space and the corresponding back radiation.

Often a **conservation of radiant energy** is tacitly assumed. This means that the radiant energy can be passed on from one greenhouse molecule to the next without much dissipation.

This condition is not fulfilled. **In molecular spectroscopy there is no "conservation of radiation energy"**. As mentioned above, a molecule excited by the absorption of radiation energy can transfer the energy, absorbed from the radiation, to other gas molecules as heat energy, instead of releasing the energy again as radiation.

In air, greenhouse gas molecules are very strongly diluted with nitrogen and oxygen. This means that when excited greenhouse gas molecules pass on their excitation energy to their collision partner, the excitation energy is practically always transferred to nitrogen or oxygen molecules. Nitrogen and oxygen molecules cannot emit the absorbed energy as IR radiation. They convert the energy gained from the collision with the excited greenhouse gas molecule into kinetic energy, i.e. into heat.

In the lower layers of the atmosphere, the density of the air is quite high. The high density causes the gas molecules here to collide with each other in a very short time sequence (order of magnitude 10^{10} collisions per second and molecule). This makes the radiationless deactivation of excited states of greenhouse gas molecules the dominant process here (thermalisation, the excited states do not have enough time to re-irradiate the absorbed radiation). Accordingly, a significant back radiation of the atmosphere to the Earth's surface is not possible in the frequency range of greenhouse gas absorptions. There can therefore be no atmospheric greenhouse effect.

The heat released near the ground through thermalisation sets convection currents in motion that transport the heat to high layers of the atmosphere (thermals).

At high altitudes, the air density is very low. Greenhouse gas molecules correspondingly rarely collide with other gas molecules. Under these conditions, excited greenhouse gas molecules have considerably more time to emit their excitation energy as IR radiation. In addition, the low density of the air at high altitudes increases the range of the radiation emitted by the greenhouse gases. At high altitudes, radiation into space (**thermally excited emission**) thus becomes the dominant process. Back radiation to the Earth's surface is not possible from high altitudes because these IR frequencies are practically completely absorbed in the lower/denser regions of the atmosphere and converted into heat.

The following picture emerges for the flux of IR radiation energy emitted by the Earth's surface at the frequencies of greenhouse gas absorptions (Figure 66):
- Irradiation from the earth's surface
- Absorption by greenhouse gases and thermalisation near the ground, i.e. conversion of radiant energy into thermal energy.
- Transport of heat energy into high layers of the atmosphere by convection currents
- In high layers of the atmosphere, conversion of thermal energy into radiation energy (thermally excited emission)
- Radiation into space

Figuratively speaking, the lower, dense regions of the atmosphere act like a check valve on the radiative energy flux at the IR greenhouse gas frequencies (Figure 66).

Now let's go back to the radiative transfer equation and see what effects the above has on the solution of the radiative transfer equation for the Earth's atmosphere. Instead of solving the equation completely, we will only look at two borderline cases.

Borderline case 1: Specific emission on the wavelength λ, $j_\lambda = 0$.
In the lower layers of the atmosphere, this premise is fulfilled quite well. Due to the practically complete thermalisation of the excited vibrational states of the greenhouse gases in combination with the short range of these wavelengths, emission virtually comes to a standstill in the lower atmosphere. The radiative transfer equation simplifies here to:

$$\frac{dI_\lambda}{ds} = -\varepsilon_\lambda I_\lambda$$

One solution to this differential equation is the Lambert-Beer's law. As discussed before, this gives rise to the saturation effect and the impermeability of the lower atmosphere to "IR greenhouse frequencies".

Borderline case 2: Specific absorption/scattering of wavelength λ, $\varepsilon_\lambda I_\lambda = 0$.
At high altitudes, the air density decreases. This increases the range of the "IR greenhouse frequencies" and thermalisation becomes less important, because the gas molecules no longer collide with each other so often per unit of time. With increasing altitude, the premise $\varepsilon_\lambda I_\lambda = 0$ is better and better fulfilled and the radiative transfer equation simplifies to:

$$\frac{dI_\lambda}{ds} = j_\lambda$$

I.e. thermally excited emission becomes the dominant process at high altitudes.

Remark:
IR back radiation measurable on the Earth's surface comes from a thin layer measuring only a few metres (order of magnitude of the previously calculated range of 15µm radiation). This is also less of a "back radiation" than a thermally excited emission of the greenhouse molecules contained in the lower layers of air. At temperatures between 15°C and 30°C, about 3% to 4% of the air molecules have sufficient kinetic energy to excite the 15µm oscillation of CO_2. How to calculate this proportion is explained below.

Satellite Measurements Confirm the Impermeability of the Lower Atmosphere to 15µm Radiation

Figure 63 shows IR spectra of the Earth recorded by satellites. The dashed curves each show the radiation of a black body with the temperature indicated in the curve. I.e. if in spectrum b the Earth had no atmosphere or an atmosphere without greenhouse gases, the satellite would measure the dashed curve for 280K.

The solid zigzag line shows the IR radiation measured over the Mediterranean. In this spectrum, the so-called "atmospheric window" can be seen very clearly. In the range from 8 to 14µm, the atmosphere is practically completely permeable to IR radiation. Only the ozone layer absorbs in this range. The radiation here follows a "blackbody radiation" that corresponds to the temperature of the radiating Earth's surface. In spectrum b, this would correspond to a temperature of the Mediterranean Sea of 7 to 10°C.

In spectra a and b, a very clear CO_2 absorption can be seen. The really interesting thing about this band is that it shows a CO_2 emission band in its centre at 15µm. The base of this emission band lies on the 220K line of the blackbody. This means that at high altitudes at a temperature of about -50°C (220K) CO_2 radiates into space (thermally excited emission).

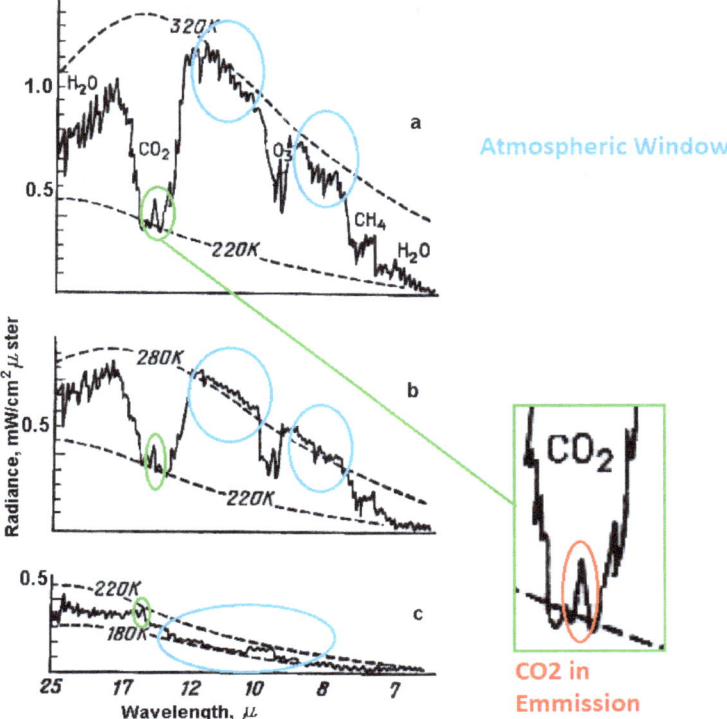

Figure 64: a - Desert Sahara; b - Mediterranean Sea; c - Antarctic Region Source of spectrum: Dr. Fred Ortenberg OZONE: SPACE VISION (Space monitoring of Earth Atmospheric Ozone) Haifa, 2002, Spectrum of Earth Thermal IR-radiance recorded from space: https://www.researchgate.net/figure/Spectrum-of-Earth-Thermal-IR-radiance-recorded-from-space-a-Desert-Sahara_fig2_291164378

Spectrum c was recorded over Antarctica. Here the Earth's surface is significantly colder than the upper part of the troposphere. Therefore, no CO_2 absorption band is visible. But if you look closely, you can see CO_2 in emission. This is a situation that should not exist if the radiative transfer equation were to describe reality.

IR radiation that cannot escape through the "atmospheric window" is practically completely converted into heat in the lower atmospheric layers. At high altitudes, it is therefore no longer available to excite greenhouse gas molecules.
At these altitudes, the greenhouse gas molecules obtain their excitation energy in collisions with other gas molecules. This means that at high altitudes the greenhouse gas molecules absorb heat energy from the atmosphere and radiate it into space as thermal radiation. **Greenhouse gases thus intensify the cooling of the atmosphere at high altitudes.**

After this excursion into spectroscopy, I would like to briefly summarise the essentials.

Saturation and thermalisation near the ground: Outside the so-called atmospheric window (wavelength range from approx. 8 to 14μm), the thermal radiation radiated from the Earth's surface has only quite short ranges in the lower atmosphere. In air layers close to the ground, the density of the air is quite high and the greenhouse gases are strongly diluted with non-IR-active gases (N_2, O_2, mixing ratio: 1 CO_2 molecule to approx. 2500 N_2 and O_2 molecules). As a result, greenhouse gas molecules collide here more or less exclusively with non-IR-active gas molecules (order of magnitude 10^{10} collisions per sec and molecule). Greenhouse gases in the excited state are therefore deactivated during collision events with nitrogen or oxygen molecules, before they find the time to emit IR radiation. The IR radiation absorbed here by greenhouse gases is practically completely converted into heat. Therefore a pronounced back radiation of greenhouse gas IR frequencies from the lower atmosphere is not possible. Here **thermalisation is the death of the greenhouse effect**, so to speak.

Heat transport by convection: From the air layers near the ground to the boundary of the troposphere, heat transport is predominantly by convection (Figure 65 and 66).

Thermally excited emission in the upper region of the troposphere: Due to the low air density at high altitudes, the emission of thermal radiation becomes more important than non-radiative deactivation. Greenhouse gases intensify the cooling of the atmosphere here (Figure 65 and 66).

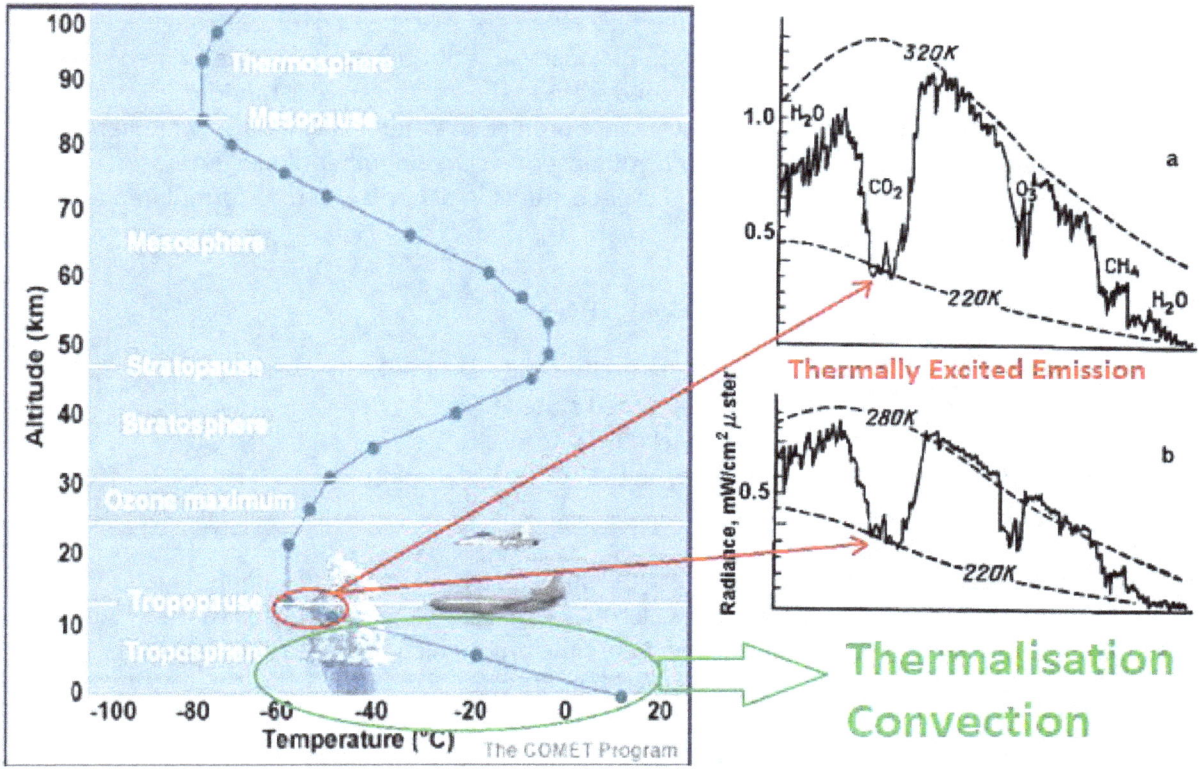

Figure 65: Thermalisation and heat transport by convection in the lower part of the atmosphere, marked in green. Radiation into space at the upper interface of the troposphere (thermally excited emission) marked in red.
The left part of the graphic taken from: https://nabilaandya98.wordpress.com/2013/05/29/science-project-earths-atmosphere/

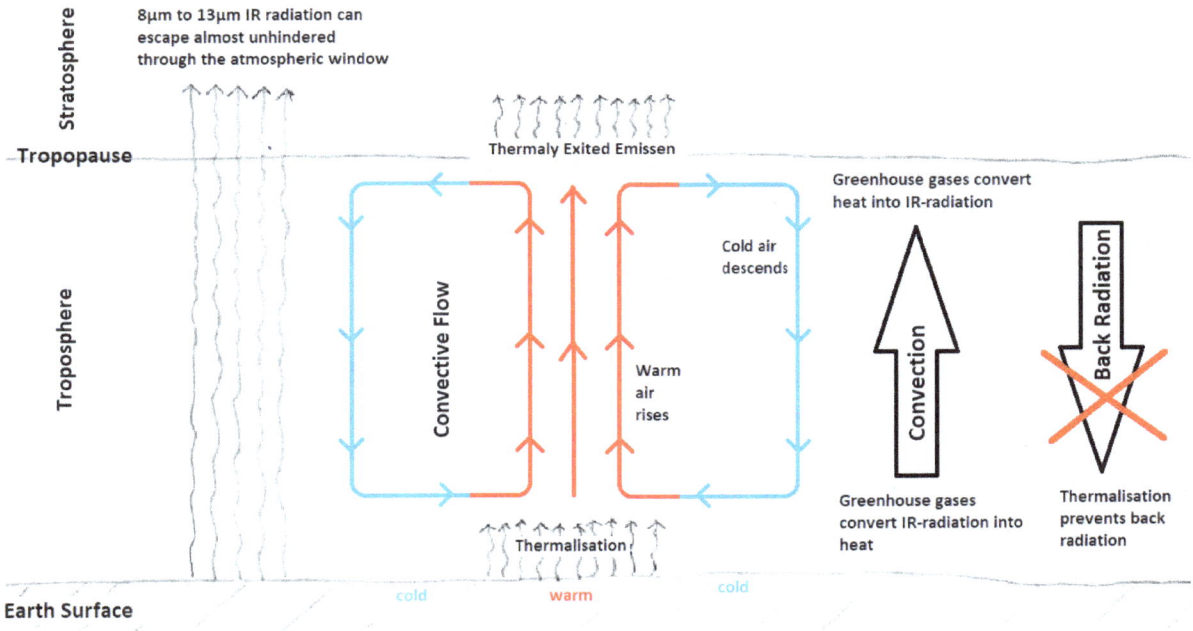

Figure 66: Heat transport in the troposphere by convection currents and thermally excited emission at the boundary of the troposphere

In order not to disturb the flow of the argumentation, I have so far only dealt with the half-life of the excited state of the CO_2 bending oscillation (01^11), the thermalisation of this state and the thermally excited emission. I now want to describe these processes in more concrete terms.

Stability of the excited oscillation state (01^10) of the CO_2-Molecuel

Similar to the radioactive decay of unstable atomic nuclei, it is not the case that the excited state has a specific lifetime. The transition from the excited state to the ground state is random. This means that if you observe a single excited state, you cannot predict whether it will radiate immediately or continue to oscillate for a few seconds. Only if you observe a large number of excited states, you will recognise a regularity in the decay rate of the excited states. A decay law can be formulated for the excited states (01^10):

$$N_i(t) = N_0 \, e^{-K_d t}$$

With: $N_i(t)$: Number of CO_2 molecules in the excited state at time t.
N_0 : Number of CO_2 molecules in the excited state at time t = 0.
$K_d = 1.542 s^{-1}$: Decay constant (HITRAN database)
https://www.spectralcalc.com/spectral_browser/db_data.php
t : Time [s]

According to this decay law, the number of excited states $N_i(t)$ will never fall completely to zero. This means that one cannot specify a time after which all excited states have returned to the ground state. Therefore, one makes do with specifying the time after which half of the excited states have returned to the ground state. This is the so-called half-life $t_{1/2}$.

When the half-life has expired, the following applies: $\frac{N_i(t)}{N_0} = \frac{1}{2} = e^{-K_d t_{1/2}}$

Dissolve to $t_{1/2}$:

$$\ln(1/2) = -K_d \, t_{1/2}$$

$$t_{1/2} = \frac{\ln(2)}{K_d} = \frac{\ln(2)}{1{,}542 \, s^{-1}} \approx 0{,}45s$$

Thermalisation of 15µm IR radiation in the atmosphere

Because CO_2 is a popular LASER medium, this process has been studied very thoroughly. The figures used in the following are from the following publication:
Thevibrationaldeactivation ofthe(0001)and(0110)Modes ofCO2measureddown to140 K by Siddles, Wilson, Simpson, Chemical Physics 189 (1994) 779-91

At 300K (room temperature) and atmospheric pressure, a CO_2 molecule suffers approx. $7 * 10^9 \approx 10^{10}$ collisions with other gas molecules. The lifetime of the excited 15µm vibrational state, i.e. the time that elapses between absorption and emission of a photon, is of the order of one second. This numerical ratio suggests that under these conditions a CO_2 molecule has hardly any chance of re-emitting a previously absorbed photon.

At atmospheric pressure and 295K (approx. 22°C), the following thermalisation rates ("quenching rates") were determined for the non-radiative deactivation of the excited state of the 15µm absoption of CO_2 in nitrogen and oxygen.

$$K_{N_2} = 5{,}5 * 10^{-15} \frac{cm^3}{Molecules * sec}$$

$$K_{O_2} = 3{,}1 * 10^{-15} \frac{cm^3}{Molecules * sec}$$

For air, a mixture of approx. 20% oxygen and 80% nitrogen, we therefore estimate the thermalisation rate:

$$K_{Air} \approx (0{,}2 * 3{,}1 + 0{,}8 * 5{,}5) * 10^{-15} \frac{cm^3}{Molekules * sec}$$

$$K_{Air} \approx 5 * 10^{-15} \frac{cm^3}{Molekules * sec}$$

Because it is not easy to imagine something "tangible" under this number, we now calculate how often per second an excited CO_2 molecule (in the 15μm absorption) is deactivated without radiation by colliding with air molecules. To do this, we multiply K_{Luft} by the Lohschmitt number, which indicates how many air molecules are contained in 1cm³ under normal conditions.

$$K_{Air} * N_{Lohschmitt} \approx 5 * 10^{-15} \frac{cm^3}{Molecules * sec} * 2{,}5 * 10^{19} \frac{Molecules}{cm^3} \approx 13{,}75 * 10^4 \frac{1}{sec}$$

$$K_{Air} * N_{Lohschmitt} \approx 10^5 \frac{1}{sec} \approx Number\ of\ non\ radiative\ deactivations\ per\ molecule\ and\ sec$$

Since the lifetime of the excited state of the 15μm absorption of CO_2 is of the order of one second, only one photon out of about 100,000 absorbed photons will be re-emitted. If one is careful and calculates with a (01^10) lifetime of about 0.1 seconds, about 1/10,000 of the excited states will have the opportunity to emit a photon before it is deactivated by collision with other gas molecules.

For other greenhouse gases such as water vapour, methane, SF_6, ... the situation is similar. Since these gas molecules are also hit by air molecules approx. 10^{10} times per second, it can be assumed that thermalisation near the ground is the dominant process for these gases as well.

In high altitudes, the proportion of emitted photons is higher due to the lower air density (fewer collisions events per time unit and molecule).

Note on thermally excited emission

Some readers are probably surprised that at the outer boundary of the troposphere at a temperature of about -50°C (220K) the air molecules have enough kinetic energy to excite the 15μm bending vibration during impact events with CO_2 molecules. I would therefore like to address this process briefly.

As described before, gas molecules move criss-cross through the space available to them. At high temperatures they move faster on average and at low temperatures slower. The Maxwell-Boltzmann distribution indicates the probability with which gas molecules have a certain velocity at a given temperature. Figure 67 shows the Maxwell-Boltzmann distribution for nitrogen at 0°C, 100°C and 1000°C. It can be seen that at high temperatures the maximum of the distribution function shifts towards high velocities. But even at low temperatures, there is still a small proportion of fast molecules in the so-called "tail" of the Maxwell-Boltzmann distribution. This small proportion of fast gas molecules makes it possible that even at low temperatures processes can take place that require quite high excitation energy.

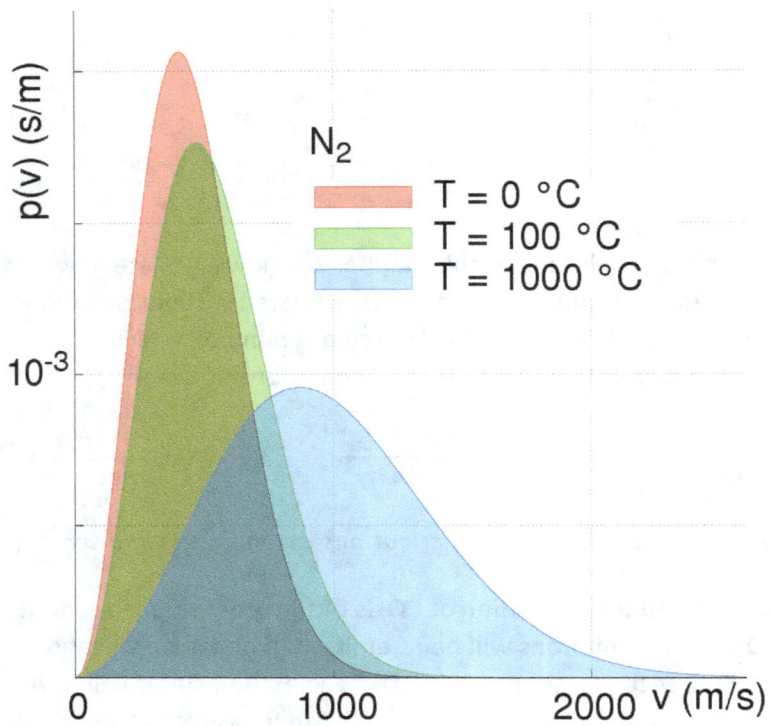

Figure 67: Maxwell-Boltzmann distribution for nitrogen at 0°C, 100°C and 1000°C, source: https://commons.wikimedia.org/w/index.php?curid=26829272

The proportion of gas molecules whose kinetic energy is greater than the excitation energy E_i is calculated as follows.

$$\frac{N_i}{N} = e^{-\frac{E_i}{kT}} = e^{-\frac{h\nu}{kT}}$$

For the excitation of the 15μm oscillation of the CO_2, this results at 220K:

$$\nu = \frac{c}{\lambda} = \frac{3*10^8 \frac{m}{s}}{15*10^{-6}m} = 2*10^{13}\frac{1}{s}$$

$$\frac{N_i}{N} = e^{-\frac{h\nu}{kT}} = e^{-\frac{6{,}626*10^{-34}Js*2*10^{13}\frac{1}{s}}{1{,}38*10^{-23}\frac{J}{K}*220K}} = 0{,}0127 \approx 1{,}3\%$$

I.e. about one percent of the air molecules have enough kinetic energy to excite the 15μm oscillation of CO_2 at the outer boundary of the troposphere at approx. -50°C.

Correspondingly, at air temperatures of 15°C to 30°C, there is a proportion of 3% to 4% of the air molecules whose kinetic energy is high enough to excite the 15μm vibration of the CO2 during collision events with CO_2 molecules.

Thermally excited emission occurs naturally everywhere in the atmosphere. At high altitudes, the emitted IR radiation can escape into space. In the denser parts of the atmosphere, the emitted IR radiation is absorbed by greenhouse gases and thermalised after a short path through the atmosphere. Close to the ground (10 to 20 m height), a small part of the radiation can reach the Earth's surface and can be measured there as very weak back radiation (much too little to cause a 33°C greenhouse effect).

Experiments to "Proof" of the Greenhouse Effect

Now that we have an idea of how the atmospheric greenhouse effect is **supposed to** work and know why this effect cannot exist, I would like to briefly discuss a "laboratory experiment to prove the atmospheric greenhouse effect" that was often presented.

This experiment demonstrates that a glass container filled with CO_2 heats up faster than its counterpart filled with air when both containers are irradiated with the same IR source (Figure 68).

After the above, this will not surprise the reader. CO_2 gas absorbs IR radiation more strongly than air and therefore heats up faster. This effect is further enhanced by the fact that CO_2 has a much lower thermal conductivity than air. The heat absorbed directly by the thermometer is dissipated more slowly in CO_2 than in air, which further accelerates the heating of the thermometer in the CO_2 container (thermal conductivity CO_2: $16{,}8 * 10^{-3} \frac{W}{mK}$, thermal conductivity air: $26{,}2 * 10^{-3} \frac{W}{mK}$).

Figure 8: Source: YouTube: https://www.youtube.com/watch?v=3v-w8Cyfoq8 , Bill Nye (undergraduate degree in mechanical engineering) brand new lab coat instead of scientific methodology,

In principle, this is the same as comparing the heating of a transparent glass plate with its matt black painted counterpart under sun light. This experiment in no way reflects the conditions in the atmosphere.

What remains of the Greenhouse effect?

We have got to the root of the effect, so to speak, and shown that the method used to determine the atmospheric greenhouse effect (of 33°C) is completely nonsensical and that the effect practically disappears when more realistic calculation methods are used.

Even at the molecular level, the mechanism of this effect (back radiation, radiation transport equation) is not comprehensible. The explanations put forward here by modern climate science are in direct contradiction to basic textbook knowledge of molecular spectroscopy, which has many technical applications (e.g. CO_2 lasers, "infrared homing wapons", ...).

In my opinion, this clearly proves that the atmospheric greenhouse effect does not exist.

For a really fundamental critique of the atmospheric greenhouse effect, I would like to refer again to the article by Gerlich and Tscheuschner.

"Falsifcation Of The Atmospheric CO2 Greenhouse Effects Within The Frame Of Physics" Version 4.0 (January 6, 2009) Prof. Dr. Gerhard Gerlich, Dr. Ralf D. Tscheuschner

Understanding these interrelations and the working of the atmosphere boosts the climate denier's self-confidence. But against the background, that mankind has no noticeable influence on the atmospheric CO_2 concentration, all this theorising is only of academic interest.

Climate Models

Climate model prediction vs. reality

I will only discuss the climate models very superficially. In principle, all you need to know is that these models make chronically wrong predictions. As can be seen in Figure 69, the predictions of the models deviate very strongly from the data determined by balloon and satellite measurements.

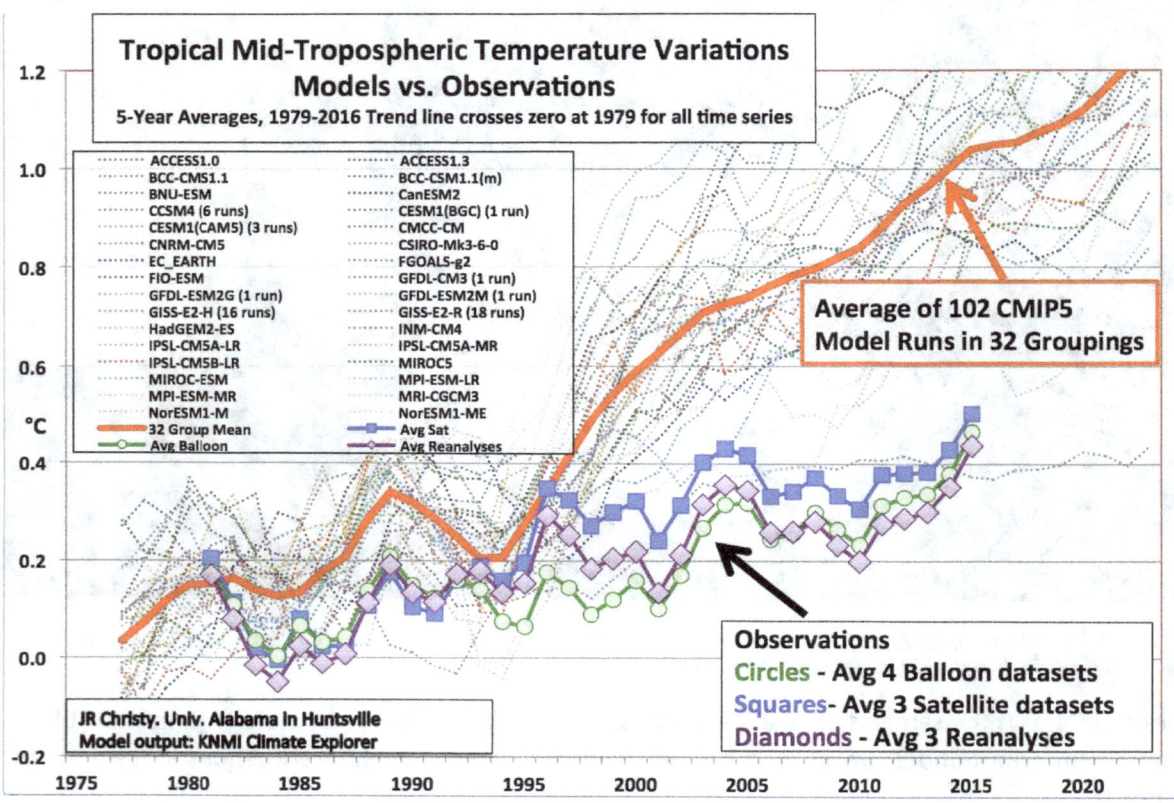

Figure 69: Comparison of temperature predictions of different climate models against actual data measured by balloon and satellite. Source: JR Christy Univ. Alabama in Huntsville:

It is also interesting to note that there are a large number of "official" models, some of which make very different predictions and thus falsify each other.

Even more interesting is the fact that one does not select the model that makes the best predictions and discards the others as wrong, but somehow averages them out quite democratically.

The IPCC is also aware of this problem. The IPCC's Third Assessment Report (2001) clearly points out that it is not possible to predict future climate states with computer models (see Figure 70).

Advancing Our Understanding

radiative forcings. This allows ensembles of model results to be constructed (see Chapter 9, Section 9.3; see also the end of Chapter 7, Section 7.1.3 for an interesting question about ensemble formation).

In sum, a strategy must recognise what is possible. ==In climate research and modelling, we should recognise that we are dealing with a coupled non-linear chaotic system, and therefore that the long-term prediction of future climate states is not possible.== The most we can expect to achieve is the prediction of the probability distribution of the system's future possible states by the generation of ensembles of model solutions. This reduces climate change to the discernment of significant differences in the statistics of such ensembles. The generation of such model ensembles will require the dedication of greatly increased computer resources and the application of new methods of model diagnosis. Addressing adequately the statistical nature of climate is computationally intensive, but such statistical information is essential.

Figure 70: Source: IPCC; Third Assessment Report (2001), Chapter 14.2.2.2. page 774, https://www.ipcc.ch/site/assets/uploads/2018/03/WGI_TAR_full_report.pdf

The models are always designed to confirm the hypothesis of man-made global warming. Models that do not support this hypothesis are not financially supported.

Apart from these financial constraints and the faulty physics and chemistry built into the models, the available computing power is not sufficient. Turbulent flows have dimensions down to the order of milimeters. The grids used in the models are on the order of 100km.

Fake climate data

The problems of the models are solved by slightly editing the data from the measuring stations. While the satellite and balloon data have observed almost no warming since 1998, ground stations succeed in measuring an increase in temperature (see Figure 71).

FAQ 1.2 (continued)

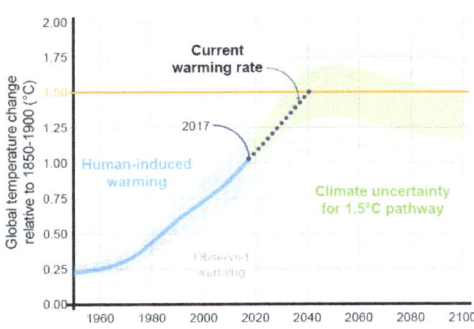

FAQ 1.2, Figure 1 | Human-induced warming reached approximately 1°C above pre-industrial levels in 2017. At the present rate, global temperatures would reach 1.5°C around 2040. Stylized 1.5°C pathway shown here involves emission reductions beginning immediately, and CO_2 emissions reaching zero by 2055.

Figure 71: Source: IPCC Report 2018, with "fudged" warming rate.

The temperature increase measured by the ground stations (Figure 71) also always correlates nicely with the increase in CO_2 concentration (Figure 72). This correlation is not visible in the more reliable satellite data (Figure 72 and 73).

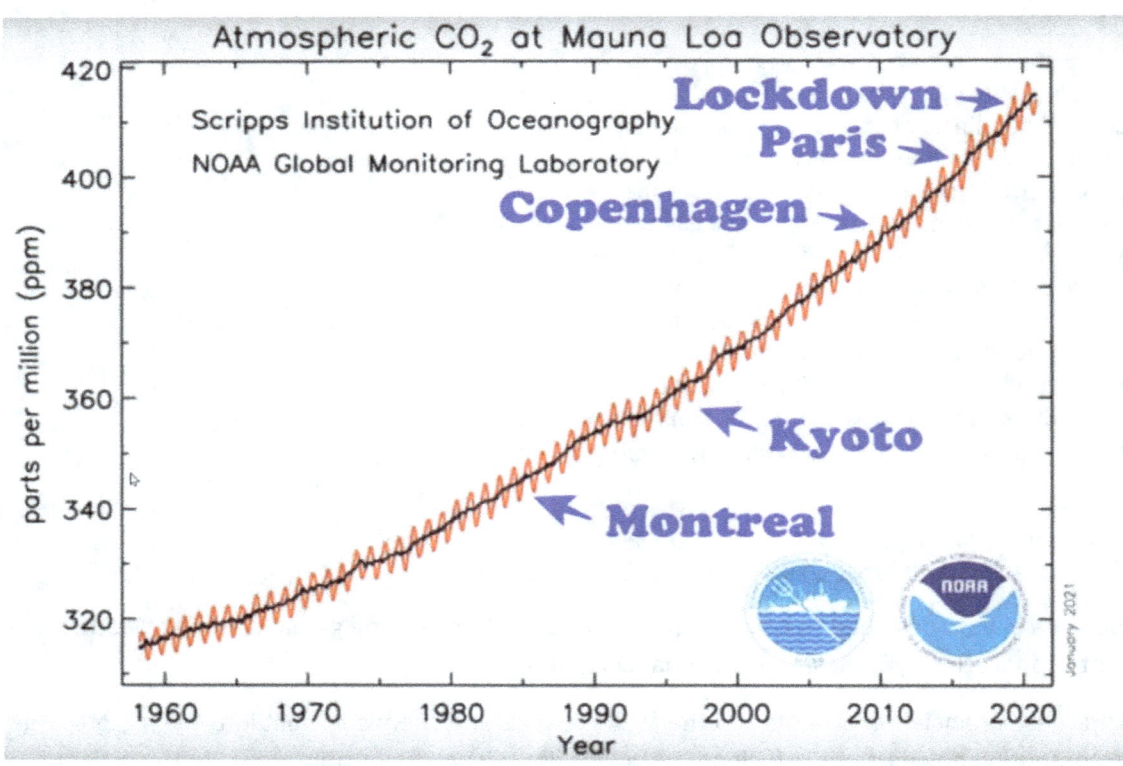

Figure 72: CO2 increase measured on Mauna Loa, source: Tony Heller (https://www.youtube.com/watch?v=OCTwukaXDqw&feature=push-u-sub&attr_tag=C6hi4B-VdM0IeKbI%3A6)

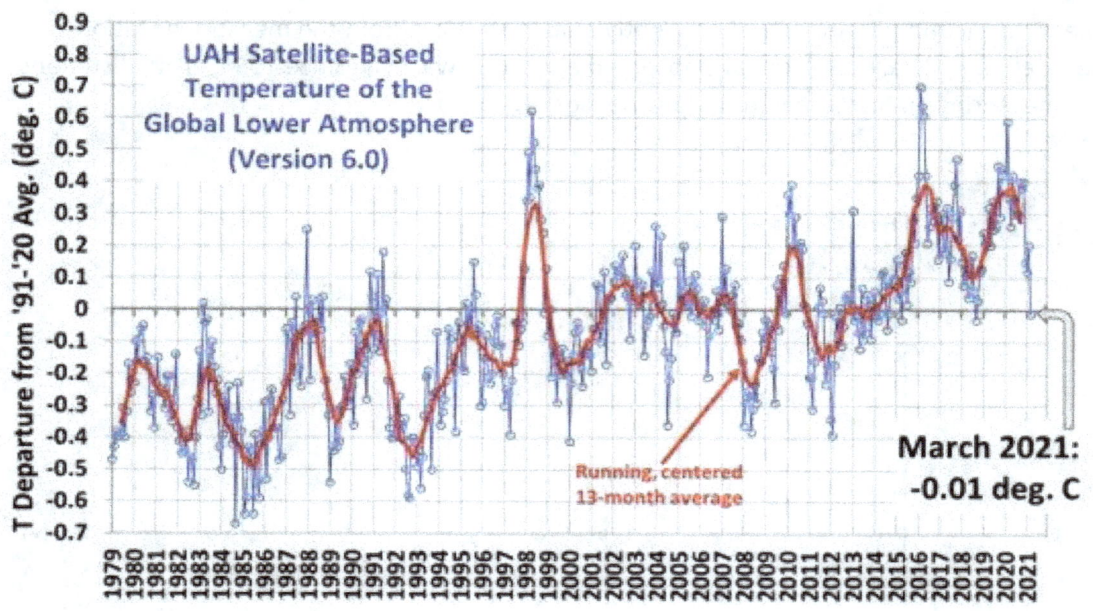

Figure 73: "Pause 1998 to 2015", Source: Roy Spencer, https://www.drroyspencer.com/

The satellite data from the Univ. of Alabama in Huntsville (Figure 72) clearly show that the hypothesis of global warming caused by the increase in CO_2 concentration in the atmosphere must be wrong. There is no correlation between CO_2 increase and global lower atmosphere temperature.

During the so-called **pause** (the period from 1998 to about 2015), global warming simply took a break, while CO_2 concentrations continued to rise unchanged (Figures 71 and 72). Even in the last four years, not much has happened in the satellite data. As already mentioned, this pause was not as pronounced in the ground station data as in the satellite and weather balloon data, which are more difficult to fake.

The attentive observer will not have failed to notice that the "pause" has made itself felt in our mainstream media. The "wording" was changed on the occasion of the "pause". For the time being, "man-made global warming" has become "man-made climate change". After the USA had to experience a series of extremely cold (up to -45°C) winters and Trump wished for good old man-made global warming back on Twitter, the preferred term was "climate disruption". Recently, the term "climate crisis" was introduced. This means something like: No more discussions. Now action is being taken.

Global warming by the greenhouse effect is still the theoretical basis of the whole scam, but one senses that officials want to get away from these terms. The simple explanation for this "new speech" is that the atmospheric greenhouse effect simply does not exist and that it is becoming increasingly difficult to hide this simple truth from the public.

As already noted, weather data has been massively faked in recent years, including in the archives, to fit the hypothesis of man-made global warming.

One has lowered in the archives especially the temperature records of the 1930s, which were warmer than the present in many places ("Dust Bowl" in N America). The temperature data of the recent past has been raised somewhat (see Figure 74). Tony Heller is intensively researching and publicising this fraud. Details are well documented on his website https://realclimatescience.com/ .

This manipulation is usually called "correction" or "homogenisation" on flimsy grounds.

The number of weather stations has also been reduced over the years, with preference given to stations with low average temperatures.

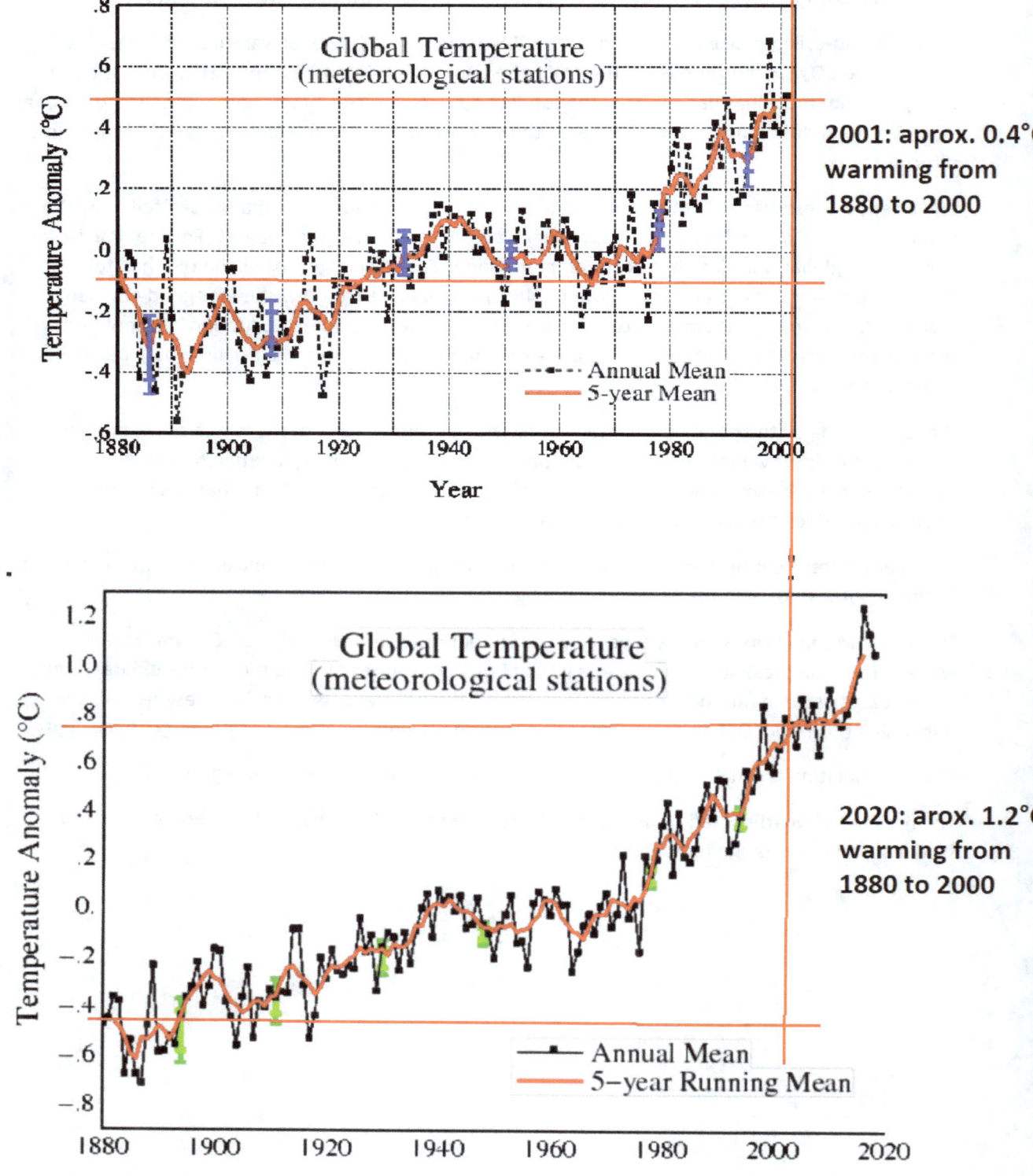

Figure 74: in 2001, NASA reports a warming of 0.5°C for the period from 1880 to 2000. In 2015, NASA reports about twice the warming for the same period. Source: Tony Heller https://realclimatescience.com/2020/01/alterations-to-giss-surface-temperatures-from-2001-to-2015/

Now that we have addressed the issue of climate data manipulation, it is worthwhile to go into a few basic concepts of climate data analysis.

Data analyses, development of models, predictions

As we have seen so far, the selection of climate data and the reconstruction of historical/prehistorical climate data open up many possibilities for cheating. When it comes to making predictions about the future of climate on the basis of existing data, there is also a lot of room for "interpretation".

The analysis of climate data is a very extensive topic that goes beyond the scope of this text. In the following, I only want to give the reader a rough idea of the fact that there are various methods of data analysis, that lead to very different results. What is said here is certainly not the last word in wisdom and is also not always mathematically correct. It is merely intended to arouse the reader's interest in this subject.

The aim of data analysis is to identify the mechanism by which the observed values are generated. Ideally, one finds a mathematical relationship between the points at which one has observed/measured and the respective measurement results. Usually this is some equation/formula of the form Y = f(X). Which assigns a value Y to each measuring point X. If you have found a formula that describes the observed values very well, you can make predictions about how the observed values will behave in the future.

When you look at a data set for the first time, in most cases you cannot see any order in the data. A real genius may guess a suitable formula. The average scientist however will simply try a few known data analysis methods until he finds one that gives useful results.

Because this sounds complicated, I will demonstrate how to do this with a simple example. I have created a small sample data set in Excel (Figure 75). The first column contains a timescale, measured in years. In the second column, temperatures, measured in °C, are assigned to each point in time. In order to put the data in a somewhat clearer form, I enter them in a coordinate system. The temperature is plotted on the vertical axis. The time is plotted on the horizontal axis.

Zeit [Jahre]	Temperatur [°C]
100	7,5
110	9,8
120	12,9
130	5,3
140	14,9
150	6,4
160	11,1
170	11,7
180	6,0
190	15,0
200	5,6
210	12,3
220	10,4
230	6,9
240	14,7
250	5,1
260	13,4
270	9,1
280	8,1
290	14,1
300	5,0
310	14,3
320	7,9
330	9,3
340	13,3
350	5,2
360	14,8
370	6,7
380	10,7
390	12,1

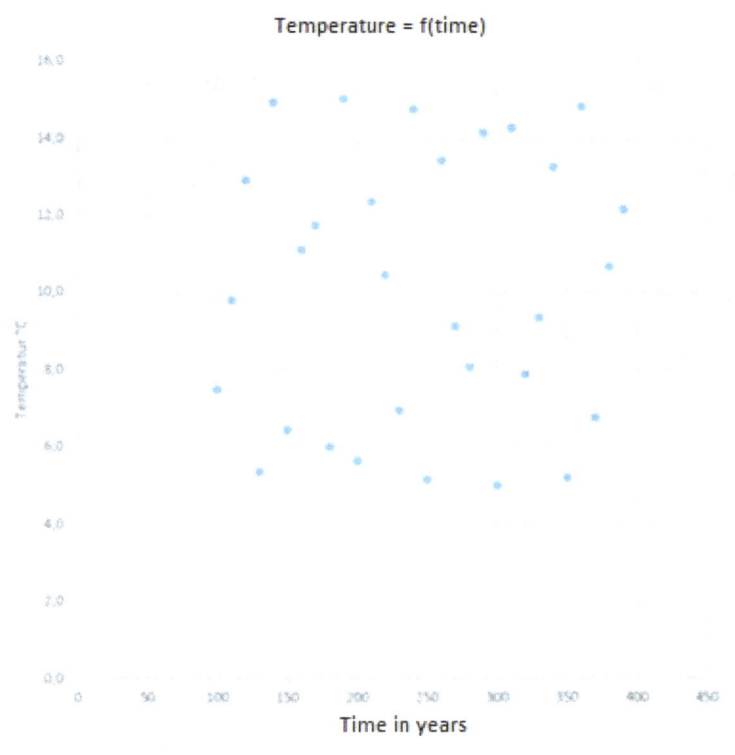

Figure 75: Example data set

This is already much clearer than the table. But we cannot yet recognise any regularity. If one were to make a prediction without much effort as to how the temperature will continue to behave after the year 390, for which one still has measured values, one would say "It continues like this". If you want to put this statement into a mathematically manageable form, you place a straight line between the points so that it passes as close as possible to all points (a so-called regression line).

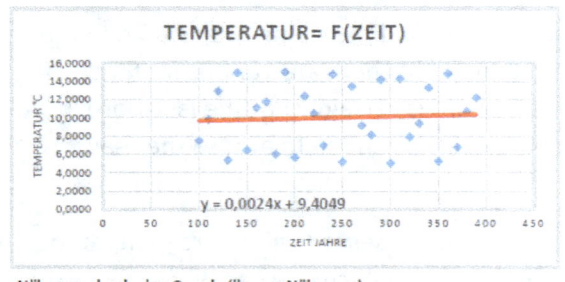 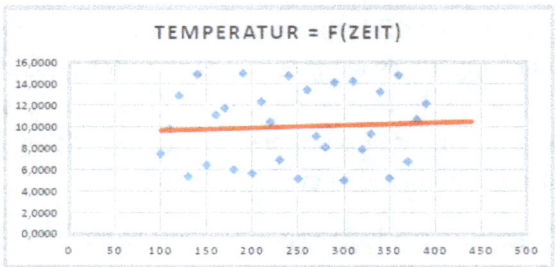

Näherung durch eine Gerade (lineare Näherung)   Vorhersage gemäß linearer Näherung

Figure 76:

Excel takes care of this (Figure 76). You get an equation for the straight line:

$$Temperature\ °C = 9{,}4049°C + 0{,}0024\frac{°C}{Year} * Time\ [Year]$$

If you want to make a prediction for the year 500, you simply extend the straight line. That is, you insert 500 years into the equation. The result of the prediction is then 10.6°C. If nothing unexpected happens, the error of this prediction will be about as large as the deviations of the measuring points from the straight line in the area that was previously measured (i.e. not bad at all). The approximation of data with the help of a straight line is called linear approximation. Prediction by this method is called linear extrapolation.

A straight line equation generally has the form:

$$Y = a + b * X$$

In our example, Excel gave me the value 9.4049 for a and the value 0.0024 for b.

If you want to draw a line between the points that passes closer to the measuring points, you can extend the straight line equation to a polynomial. This then looks like this:

$$Y = a + b * X + c * X^2 + d * X^3 + e * X^4 + and\ so\ on...$$

This extension of the straight line equation ensures that the line passes closer to the measuring points than the simple straight line. Don't worry, here too Excel calculates the best values for a, b, c, d, e … for us. This is what it looks like (Figure 77):

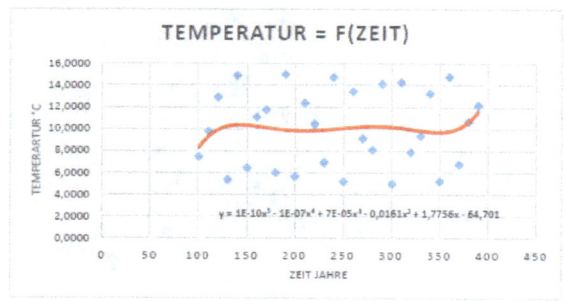 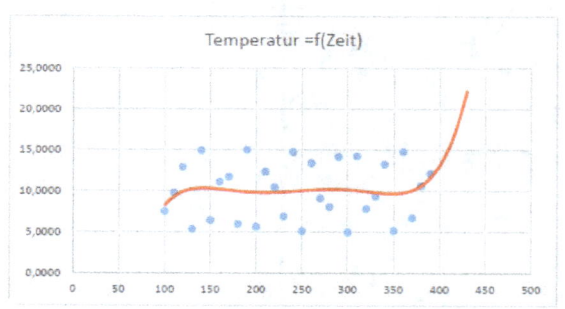

Näherung durch Potenzreihe   Vorhersage gemäß Potenzreihe

Figure 77:

Since the last data of our example are rather in the upper range of the measured values, the polynomial (4th degree) used as an approximation runs upwards at the end. If you use this polynomial to make a prediction, you get very high values that do not really fit into the picture with this data set.

The fact that my polynomial, generated in a few seconds with Excel, looks not that different to the official IPCC predictions of global warming is no coincidence (see Figure 78, Hockey Stick Graph, Michael Man). If one is looking for clear trends in any data, the polynomial is a good choice as an approximation.

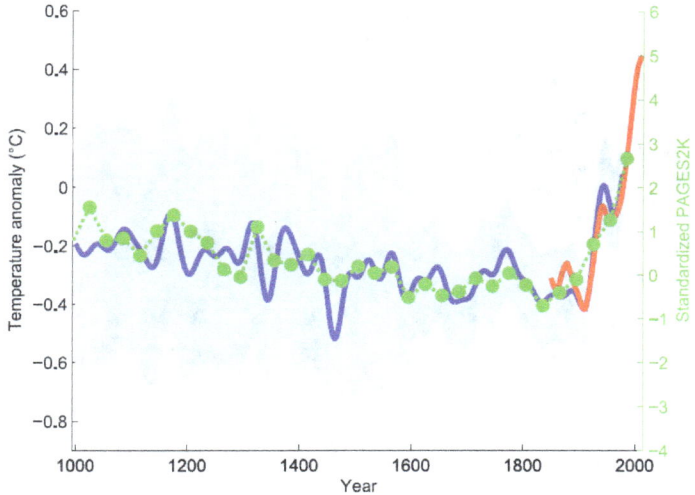

Figure 78: Hockey stick graphs, Michael Man, source: https://en.wikipedia.org/wiki/Hockey_stick_graph

In the 1970s, a clear cooling trend was observed, starting from the 1940s (Figure 79).

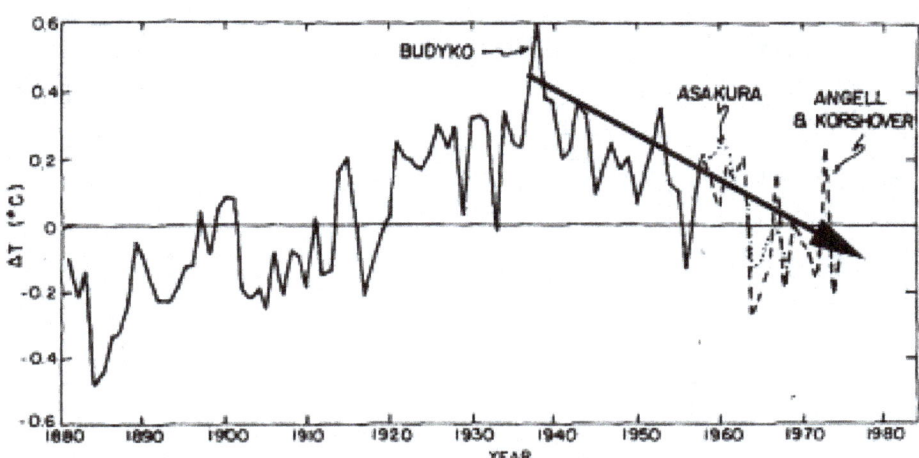

Figure 79: Source: https://joannenova.com.au/2016/09/history-rewritten-global-cooling-from-1940-1970-an-83-consensus-285-papers-being-erased/

Accordingly, "climate science and quality media" warned us of the imminent new ice age (Figure 80).

Figure 80; Source: Time Magazine Dec. 3, 1973 and Apr. 3, 2006
http://content.time.com/time/covers/0,16641,19731203,00.html
http://content.time.com/time/covers/0,16641,20060403,00.html

I know that even the suggestion that the climate models could in principle be something like a polynomial will bring me harsh criticism. But what counts in the end is the result. And that looks like a somewhat extended polynomial.

Next, we apply Fourier analysis to the sample data set. Fourier analysis is particularly suitable for the analysis of oscillating systems.

In nature, there are many systems that do not have a clear trend, but oscillate around some "zero line" (e.g. the annual cycle of day length). Fourier analysis was developed to analyse such systems. If you apply this analysis to data sets, you can see whether the observed values are the result of oscillations.

Applied to my small sample data set, Fourier analysis yields an interesting result (Figure 81).

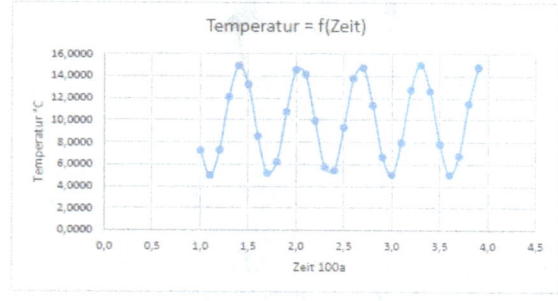
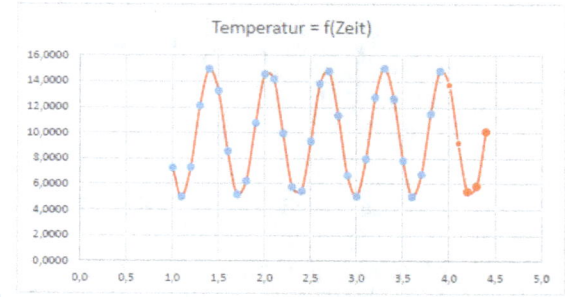

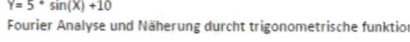

Figure 81:

The seemingly randomly arranged data points in Figure 75 can be represented here as the result of an oscillation. The oscillation that generated these data is described by the function:

$$Y = 5 * sin(X) + 10$$

The predictions that this data analysis allows are dramatically different from the predictions that the polynomial makes. If one wants to decide whether the polynomial or the sinusoidal function better describes the observed process, one will observe the system for a few more years. If the temperatures decrease over time, one can assume that the sine function describes the system quite well. If, on the other hand, one observes rising temperatures, then the polynomial seems to describe the system better. Against this background, the record-breaking temperatures repeatedly announced in our main stream media are to be understood. In addition, one can also compare how well the formulas obtained from data analysis fit the values measured in the past. In this comparison, the sine function is superior to all other approximation formulae and seems to describe the system correctly.

Our world is undoubtedly a system governed by repetitive processes. The moon moves around the earth. The Earth, together with other planets, moves around the Sun. The solar system moves around the centre of the Milky Way. The activity of the sun changes regularly, ... Seasons, ice ages, warm ages, ... come and go. Fourier analysis should therefore be the tool of choice for analysing weather and climate data. Horst-Joachim Lüdecke and Carl-Otto Weiss have performed an appropriate data analysis (*The Open Atmospheric Science Journal*, 2017, *11*, 44-53). From various temperature datasets, they compiled a global temperature dataset covering the period from 1 AD to 2015. They subjected this data set to a Fourier analysis. It turns out that the course of this temperature data set can be approximated very well by a superposition of four sine functions. These sine functions have periods of approx. 1003years, 463years, 188years and 65years (Figure 82).

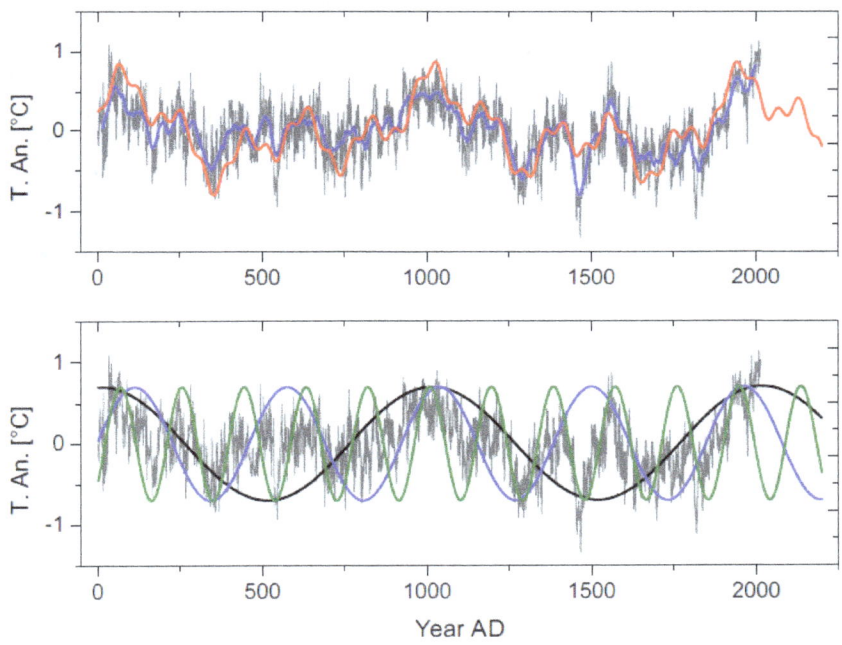

Fig. (3). (Color online) Upper panel: Global record G7 (grey), running 31-year average of G7 (blue), sine representation of G7 with three sine functions of the periods 1003, 463, and 188 years (green), with four sine functions including the period ~60 years (red), continued to AD 2200. The parameters of the sine functions are given in Table 3. The Pearson correlation between the 31 year running average of G7 and the three-sine representation (green) is 0.84, for the four-sine representation (red) 0.85. Lower panel: G7 (grey) together with the sine functions of 1003, 463, and 188 - year periods continued until AD 2200 (equal sine amplitudes for clarity).

Figure 82: Source:
https://www.researchgate.net/publication/318366114_Harmonic_Analysis_of_Worldwide_Temperature_Proxies_for_2000_Years

If this analysis is correct, the warm climate of recent years is explained by the fact that the overlapping climate cycles all went through their maximum at the same time. A similar situation existed at the beginning of the analysed global temperature data set (Roman Warm Period) and around the year 1000 (Medieval Climate Optimum). However, this also means that our current warm period will only be temporary. The authors therefore predict a significant cooling in the near future.

The authors point out that the climate cycles they observed fit well with the temporal course of the concentrations of the radionuclides ^{14}C and ^{10}Be. This suggests a connection between solar activity and climate events.

The idea, that world and climate events are subject to cosmic cycles, is not new. Almost all great civilisations have developed cosmologies in which world events are described as a regular sequence of cycles. In these cycles, civilisations rise and fall. The best-known examples of such cycles are the ages (yugas) of Hindu cosmology (Figure 83) and the cycles of the Mayan calendar. Ragnarok, the twilight of the gods in Norse mythology, is also a quite popular example of an apocalypse predetermined by cycles.

Against this background, the suspicion arises that astrology had its origins in a sound science which, thanks to the knowledge of these cycles, was able to give valuable advice on climate-related matters. Over time, the need to make money has corrupted this art somewhat. Parallels to modern climate science are probably not entirely coincidental.

In a world determined by cycles, one should not be surprised if one repeatedly finds biblical passages that describe our current situation amazingly well (Figure 83).

Figure 83: **Isaiah 3:12 (King James Version, Judgment on Judah and Jerusalem)** "children are their oppressors, and women rule over them".

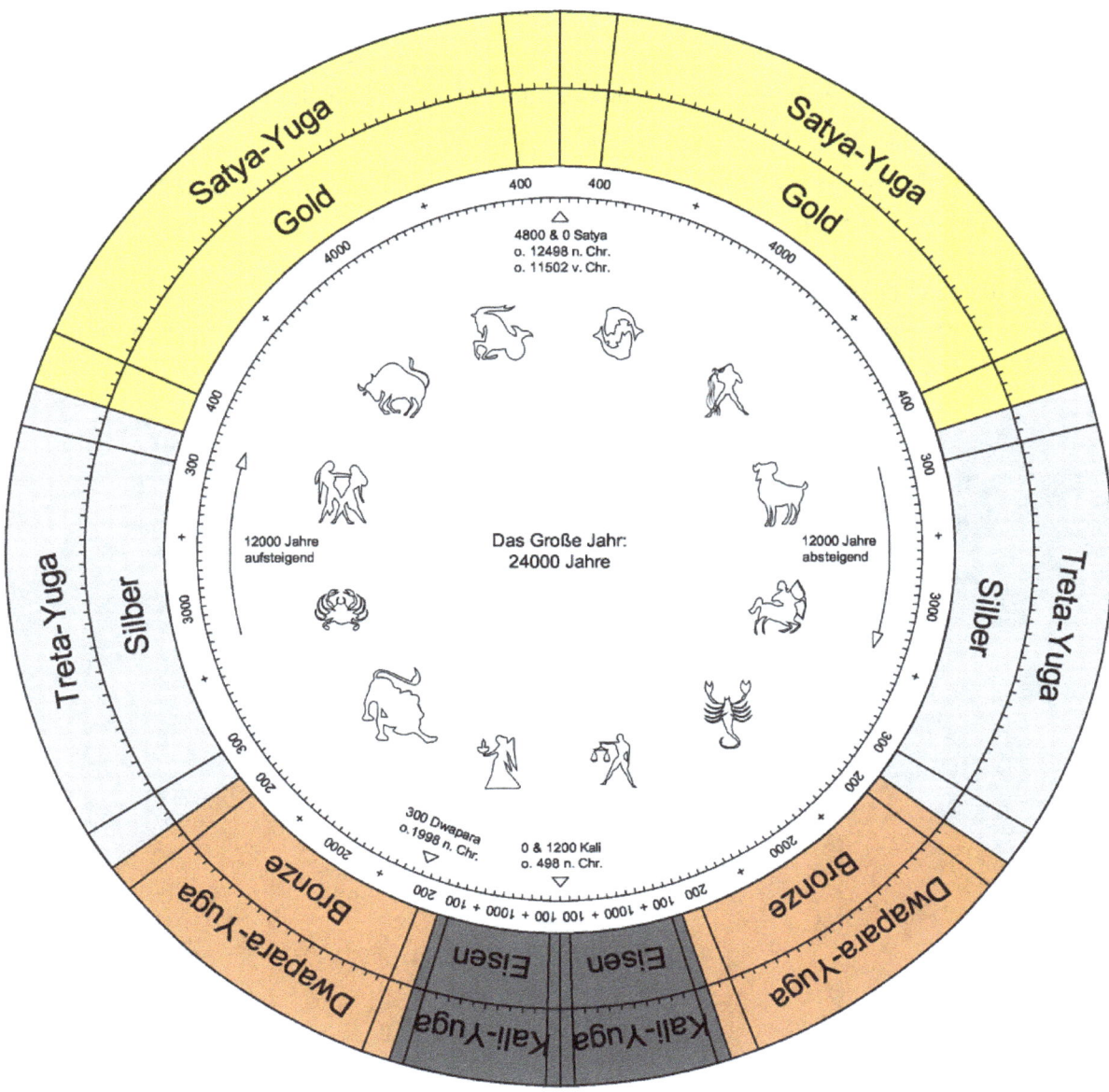

Figure 84: Source: Ingo Kappler --Inka 23:32, 8 May 2005 (UTC) - Own work, CC BY-SA 2.0 en, https://commons.wikimedia.org/w/index.php?curid=134870

Further information

For readers who want to learn more about the religious and social background of the "CO_2 cult", I would like to recommend the work of Prof. Dr. Edward Dutton. Caution. I can only recommend the works of Prof. Dutton with an explicit "trigger warning".
https://www.bitchute.com/video/KQfJmHmJopo/

For readers who like to wallow in conspiracy and intrigue, I recommend taking a closer look at the Climategate scandal. In 2009, the email communications of leading climate scientists were hacked and published. This material clearly shows that the public is being lied to and deceived on a grand scale. https://www.corbettreport.com/interview-629-tim-ball-on-climategate-3-0/
https://www.youtube.com/watch?v=LCXVZpGoDCA

A good introduction to the topic of climate cycles and their effect on world events can be found in the video "Cycles of History" by Peter Temple.

https://www.youtube.com/watch?v=MUxzW4gi8kg&list=FLOnqDHh854YsYaShFgGRrPQ&index=7&t=823s

The website of the Cycle research Institute is also of interest in this context.
https://cyclesresearchinstitute.org/

For readers who want to get ahead of the next paradigm shift and know today what will become the "new normal" in the geosciences, I would recommend James Maxlow's "Expansion Tectonics".
https://www.jamesmaxlow.com/ , https://www.youtube.com/watch?v=8qoTs7w22r4

Changes in the composition of the Earth's atmosphere appear in a completely different light against the background of this model. The much-discussed hypothesis of "peak oil" will also not be tenable with this model.

www.ingramcontent.com/pod-product-compliance
Lightning Source LLC
Chambersburg PA
CBHW081051170526
45158CB00007B/1942